Developments in
Design Methodology

Developments in Design Methodology

Edited by Nigel Cross
The Open University

JOHN WILEY & SONS
Chichester · New York · Brisbane · Toronto · Singapore

UNIVERSITY OF TOLEDO LIBRARIES

· Copyright © 1984 by John Wiley & Sons Ltd.

All rights reserved.

No part of this book may be reproduced
by any means, nor transmitted,
nor translated into a machine language
without the written permission of the publisher.

Library of Congress Cataloging in Publication Data:
Main entry under title:

· Developments in design methodology.

   Includes bibliographies and index.
   1. Design—Methodology—Addresses, essays, lectures.
   I. Cross, Nigel.
NK1525.D48   1984   745.4′01′8   84–7433

ISBN 0 471 10248 2

British Library Cataloguing in Publication Data

Design Methodology.
   1.   Design
   I.   Cross, Nigel
         745.4′01′8   NK1510

ISBN   0   471   10248   2

Typeset by Photo-Graphics, Honiton, Devon and
printed by the Pitman Press, Bath, Avon.

NK
1525
D48
1984

Contents

Introduction

It would seem to be either a very brave or foolish thing to do, to assemble a book on design methodology. There is a near-universal fear and loathing of methodology, and methodologists are reviled as impoverished creatures who merely study, rather than practise, a particular art or science. Voltaire said 'Theology is to religion what poison is to food', and there are many who would draw the same parallel between methodology and design. One critic, Christopher Alexander, has said, 'If you call it, "It's a Good Idea To Do", I like it very much; if you call it a "Method", I like it but I'm beginning to get turned off; if you call it a "Methodology", I just don't want to talk about it' (Alexander, 1971). Design methodologists found this comment particularly wounding since it came from one of the world's leading design methodologists!

Part of the distrust of methodology arises from an ambiguity in the meaning of the word. In the sense of 'a methodology' it can mean a particular, prescribed, rigid approach, of which practitioners are usually justifiably sceptical. But the sense of 'methodology' in which it will be used here is to mean the general study of method. So anyone who wishes to reflect on how they practise their particular art or science, and anyone who teaches others to practise, must draw on methodology. However, this is not to say that anyone who practises or teaches must be a methodologist. The distinction has been made by Sir Frederick Bartlett, in the context of experimental science: 'The experimenter must be able to use specific methods rigorously, but he need not be in the least concerned with methodology as a body of general principles. Outstanding "methodologists" have not themselves usually been successful experimenters' (Bartlett, 1958).

Design methodology, then, is the study of the principles, practices and procedures of design in a rather broad and general sense. Its central concern is with how designing both *is* and *might be* conducted. This concern therefore includes the study of how designers work and think; the establishment of appropriate structures for the design process; the development and application of new design methods, techniques, and procedures; and reflection on the nature and extent of design knowledge and its application to design problems.

The development of this relatively new field of design methodology has been conducted principally through means such as conferences and the publication of research papers. There is only one general textbook of design methods, by Jones (1970), and only in 1979 was there established a comprehensive, academic, international journal in the field, the journal of *Design Studies*. For students, teachers, and researchers in architecture, engineering design, industrial design, and other planning and design professions, and for newcomers generally to the field of design methodology, there has been no easy way by which they could become familiar with the field and its history of development. The important papers are scattered through many publications, in conference proceedings and so on, and many are often difficult or impossible to obtain. The purpose of this book, therefore, is to bring together in one volume a set of papers that reliably traces the development of design methodology, and comprehensively maps out the field.

Inevitably, many important papers and publications have not been included here, but the difficult decision to exclude has normally been made on the grounds of whether or not they are relatively easily available elsewhere. References to these other important publications are given with the introductions to each of the separate Parts of this book. It must be realized, however, that this is not a book of design *methods*, nor does it offer *a methodology* for design; it is intended as a reference work for the more general field of design methodology.

The book is divided into five Parts, reflecting both the main sub-divisions of the field and also to some extent its chronological development. The papers have been selected to cover the 20-year period from the first Conference on Design Methods, held in London in 1962 (Jones and Thornley, 1963), to the much larger and more wide-ranging Design Policy Conference, also held in London, in 1982 (Langdon *et al.*, 1984). These two conferences might perhaps be regarded as marking the 'birth' and the 'coming of age' of design methodology.

The conferences which have recorded the growth and development of design methodology between these two points also have been mostly held in Great Britain. The major ones were held in Birmingham in 1965 (Gregory, 1966); Manchester in 1971 (Cross, 1972); London in 1973 (unpublished); and three times in Portsmouth: in 1967 (Broadbent and Ward, 1969), in 1976 (Evans, *et al.*, 1982), and in 1980 (Jacques and Powell, 1981). Most of these conferences have been sponsored by, and held under the auspices of, the Design Research Society. In the USA, two principal conferences which should be mentioned were those held in Boston in 1968 (Moore, 1970) and in New York in 1974 (Spillers, 1974).

These selected conferences have mostly taken a catholic attitude towards design methodology and design research, although some have emphasized particular themes (such as design participation at Manchester in 1971) or particular professional areas of design (such as architectural design at Portsmouth in 1967 and Boston in 1968). The particular 'themes' of this book are reflected in the titles of its separate Parts. It does concentrate more on one professional area than others—on architecture, environmental design, and planning—partly because in that area there seems to have been the most wide-ranging and long-standing interest in all the themes of design methodology.

The papers in Part One, The Management of Design Process, are drawn from the period 1962–67. They reflect the concern of the 'design methods movement' in the early 1960s with the development of systematic procedures for the overall management of the design process, and of systematic techniques to be used within such a process. This was the period of 'systematic design' in which attempts were made to restructure the design process on the basis of the new methods and techniques of problem solving, management, and operational research which had been developed in World War 2 and in the 1950s.

It soon became realized, however, that design problems were not so amenable to systematization as had been hoped. Attention turned to trying to understand the apparent complexity of these particular kinds of problems. The papers in Part Two, The Structure of Design Problems, reflect this concern and are drawn from the period 1966–73. The major issues in this area of design methodology revolve around the 'ill-structuredness' of design problems.

Another line of approach into the complexity of design problems and the strategies for resolving them is to develop a greater understanding of how designers tackle such problems with their normal, conventional design procedures. Part Three, The Nature of Design Activity, embodies a range of methods of enquiry which have been used to investigate designer behaviour, from controlled laboratory experiments to open-ended interviews. The particular set of papers chosen to illustrate this range were all, coincidentally, published in 1979, perhaps indicating a peaking of interest in this kind of investigation in the late 1970s, although studies of designer behaviour have been made since the earliest days of design methodology.

The papers in Part Four, the Philosophy of Design Method, are drawn from the period 1972–82. They reflect the more fundamental and philosophical approach which emerged in the second decade of design methodology. This more mature and reflective approach has been able to draw upon the knowledge gained and the lessons learned in the first decade.

Finally, Part Five reviews the History of Design Methodology as told in the words of some of its leading participants. One of the surprising things in the 20 years covered by this book has been the way attitudes and opinions have changed quite dramatically. Protagonists have become antagonists, and internal debate has become internecine conflict.

It is hoped that this book might help to return the discussion and study of design methodology to an appropriate level of discourse. The way the book is structured suggests a subdivision of major interest areas which might constructively be pursued from the foundations now established. The structure also suggests that the 'movement' has progressed through four stages: *prescription* of an ideal design process (Part One), *description* of the intrinsic nature of design problems (Part Two), *observation* of the reality of design activity (Part Three), and *reflection* on the fundamental concepts of design (Part Four). Progressing through these stages might well have been an inevitable process of maturation. In any event, design methodology now seems in a much stronger condition to return to its origins, to the prescription of realistic ideals.

REFERENCES

Alexander, C. (1971), 'The state of the art in design methods', *DMG Newsletter*, **5** (3), 3–7.

Bartlett, F. C. (1958), *Thinking: an Experimental and Social Study*, Allen and Unwin, London.

Broadbent, G., and Ward, T. (eds) (1969), *Design Methods in Architecture*, Lund Humphries, London.

Cross, N. (ed.) (1972), *Design Participation*, Academy Editions, London.

Evans, D., Powell, J., and Talbot, R. (eds) (1982), *Changing Design*, Wiley, Chichester.

Gregory, S. A. (ed.) (1966), *The Design Method*, Butterworth, London.

Jacques, R., and Powell, J. (eds) (1981), *Changing Design*, Westbury House, Guildford.

Jones, J. C. (1970), *Design Methods*, Wiley, Chichester.

Jones, J. C., and Thornley, D. (eds) (1963), *Conference on Design Methods*, Pergamon, Oxford.

Langdon, R., Baynes, K., Cross, N., Gregory, S., Mallen, G., and Purcell, P. (eds) (1984), *Design Policy*, The Design Council, London.

Moore, G. T., (ed.) (1970), *Emerging Methods in Environmental Design and Planning*, MIT, Boston, Mass.

Spillers, W. R. (ed.) (1974), *Basic Questions of Design Theory*, North-Holland, Amsterdam.

Part I

The Management of Design Process

Introduction

One of the first tasks attempted by design methodologists was the development of new, systematic design procedures. In the early 1960s these new procedures began to emerge in all the different professional specialist areas of design. The four papers here in Part One represent engineering design, urban design, industrial design, and architectural design. They are drawn from the period of the early to middle 1960s, starting with two of the seminal papers—those by Jones and Alexander—given at the first 'Conference on Design Methods' in London in 1962.

Jones' 'A method of systematic design was one of the first attempts to provide a completely new way of proceeding with design. It did not, however, attempt to replace every aspect of conventional design; it was based on the recognition that intuitive and irrational aspects of thought have just as important roles to play in design as logical and systematic procedures. The method was aimed particularly 'at the area that lies between traditional methods, based on intuition and experience, on the one hand, and a rigorous mathematical or logical treatment, on the other'.

This clear intention to supplement, rather than to supplant, traditional design methods was often ignored by the early critics of systematic design procedures, who tended to assume that the 'systematic' must be the enemy of the 'intuitive'.

What Jones was proposing was a way of organizing the design process so that logical analysis and creative thought—both assumed to be necessary in design—would proceed in their own different ways. His systematic design method attempts to recognize and to separate the two ways of thinking by the use of clear, externalized procedures, rather than leaving them as internal mental battles for the designer. In particular, his Method attempts to leave the designer's mind as free as possible for random, creative ideas or insights, by providing systematic methods for keeping data, information, requirements, and so on, outside the memory.

Typical aspects of this method of systematic design therefore include keeping separate records for ideas or solutions and information or requirements; not suppressing random inputs, but storing them for later evaluation; systematically sorting the interactions between problem factors; delaying the choice of a final solution until the problem is fully explored and potential solutions are evaluated; and combining partial solutions into a whole, rather than the reverse procedure of conventional designing.

The method permits, and even enourages, random, unstructured thinking (e.g. use of a 'random list of factors' and 'creative thinking'). It provides a rational framework within which the irrational has its own space and time. The framework consists essentially of a procedure of three stages—analysis, synthesis, and evaluation—and within each stage Jones offers a variety of techniques appropriate to the main task. The three-stage process of analysis–synthesis–evaluation became widely accepted as a basic model of the systematic process of design.

Alexander's paper on 'The determination of components for an Indian village' was drawn from the research for his Doctoral thesis, and his later book, *Notes on the Synthesis of Form*. Part of this research drew upon biological analogies of organisms and environments for the relationship between design forms and their contexts. His use of the term 'environment' in this paper therefore refers to the general problem context—i.e. the design requirements of the village he is designing as a case-study. This Indian village is taken as an example of 'a city in miniature', for it is city planning or urban design which is the focus of his attention.

Because the basic components into which any artefact can be subdivided determine the essential nature of that artefact, Alexander addresses 'the general problem of finding the right physical components of a physical structure'. He seeks to conceptualize new components; to design totally new, more appropriate artefacts, structures, and systems. A radically new structure for a city, for example, cannot emerge from simply rearranging the accepted conventional components. Alexander's concern, therefore, is to find a way of formulating components which does not rely on preconceptions of what those components should be; the components should derive from a thorough analysis of the environment they have to fit. This leads him to doubt the value of most systematic design approaches (such as applications of linear programming or decision theory techniques), which take existing solution components as their starting points and therefore merely rearrange those same components.

Systematic techniques [he writes] just because they need to operate on known units, usually beg the real question of design, and so achieve little more than a second rate designer does. The fundamental change which a structure undergoes at the hands of a great designer, who is able to redistribute its functions altogether, cannot take place if its components stay the same.

Alexander's own design method therefore starts from scratch with observation of the problem context and an exhaustive listing of requirements. Then one has to decide for each pair of requirements whether or not they interact, or are dependent. 'Two requirements are dependent if whatever you do about meeting one makes it either harder or easier to meet the other, and if it is in the *nature* of the two requirements that they should be so connected, and not accidental.' The subsystems of the problem context are then derived by partitioning the linked requirements into independent subsets. Since this is a large and complex task, Alexander derived a computer method for doing this, based on graph theory. The result is the formulation of a set of subsystems of the environment; the remaining task is to design components to match the subsystems.

In the Indian village example, Alexander lists 141 basic requirements. Following his method, these are grouped into twelve independent minor subsystems, which can be combined into four major subsystems. For each subsystem he provides a diagrammatic concept for a matching component.

Alexander's work was very influential, and his method of hierarchical decomposition of a set of problem requirements was used in many other design fields.

A third influential, early contribution to design methodology was a series of articles on 'Systematic method for designers' by L. Bruce Archer published in *Design* magazine during 1963 and 1964, and later republished by the Design Council. Almost the whole text is reprinted here, except for the substantial checklist of design procedure (which ran to some 13 printed pages) and the accompanying network flow diagrams showing the relationships between the 229 activities in the checklist.

Archer's context is industrial design, and his opening comments on the nature of designing reflect this orientation. For him, design activity is based on the formulation of a prescription or model which represents the intention to create some artefact, and the activity must include some creative step. He distinguishes designing from artistic creation, musical composition, scientific discovery, and mathematical calculation. He makes it clear that systematic designing does not imply automatic designing, and also argues that rigorous analysis does not necessarily result merely in statements of the obvious—any 'obviousness' is only apparent in retrospect.

Archer's model of the design process is more complicated than many others. His complete checklist contains 229 activities, in nine major stages from 'preliminiaries' to 'winding-up'. However, the core of his model is a six-stage process: programming, data collection, analysis, synthesis, development, and communication. 'In practice', he writes, 'the stages are overlapping and often confused, with frequent returns to early stages when difficulties are encountered and obscurities found.' Thus his diagrammatic model of the design process contains many feedback loops. Simplifying even further, his model reduces to 'a creative sandwich', with a central creative phase sandwiched between more objective phases of analysis and execution.

Although he is a strong advocate of rigorous and thorough analysis, Archer recognizes that perfect and complete information is rarely available in the real world, and that it is therefore not possible to wait until an analysis has been conducted of *all* the potentially relevant data before the designer has to act. The designer's previous experience, and knowledge of case histories, is important in formulating a reasonable course of action on incomplete evidence, and in fact, 'Making a first approximation on the basis of prior experience enormously reduces the scale of the problem solving effort.' However, he also says that 'It is axiomatic that any rational method for solving design problems must offer means for arriving at decisions on the basis of evidence'. and he therefore stresses the collection and organization of information so that the designer can make sensible decisions.

The opening analytical phase of Archer's systematic design procedure essentially comprises: identifying the design goals; identifying the constraints; preparing a list of sub-problems; and rank-ordering the sub-problems. 'The result, however, is a statement of the problem, not of the answer', and the bread of objective and systematic analysis must be followed by the meat of the 'creative leap'. Archer insists that there is 'no escape' for the designer from creatively formulating his own design ideas. Creativity is part of the very essence and nature of designing: 'After all, if the solution to a problem arises automatically and inevitably from the interaction of the data, then the problem is not, by definition, a design problem.' The 'creative leap' to a potential solution is then followed by the 'donkey work' of constructing development models, which can be drawings or other analogues, representing a particular embodiment of the general solution idea.

There are several similarities between Archer's view of design and that of Luckman, in his paper 'An approach to the management of design', although the context for the latter is architectural design and the viewpoint is that of an operational researcher. Luckman emphasizes the analysis of information, requirements, and constraints, which the designer translates, with the help of experience, into potential solutions which meet the required performance characteristics of the artefact being designed. He also insists that 'some creativity or originality must enter into the process for it to be called design', and that 'if the alternative solutions can be written down by strict calculation, then the process that has taken place is not design'.

Luckman's model of the design process is based on the three-stage process of analysis–synthesis–evaluation. However, his view is not that this is a simple, complete, linear process, but that it recurs at different levels of design detail; the designer is continually cycling through analysis–synthesis–evaluation, proceeding from the more general problem levels to the more specific. At each level, Luckman perceived from his observations of architects, the components of a solution are always highly interdependent, and the designer's difficulty therefore lies in finding a compatible set. There is no guarantee that optimum sub-solutions will combine into an overall optimum solution. It was this problem that Luckman addressed with his systematic design procedure, AIDA—the analysis of interconnected decision areas.

A 'decision area' occurs where, at any level of detail, there is a range of acceptable sub-solutions to a particular sub-problem. Because the choice of a sub-solution usually influences, and is influenced by, other sub-solutions to other sub-problems at the same level, the majority of decision areas are highly interconnected. The AIDA method enables the designer to identify compatible sets of sub-solutions and so to make 'simultaneous' choices rather than sequential ones. Despite his belief in the role of creativity in design, according to Luckman, 'AIDA is really a systematic technique for the synthesis stage'. It expands the number of solutions to be considered in the evaluation stage, instead of the conventional reliance on considering only a few potential solutions—or even simply the first solution to emerge.

If we review these four early contributions to the development of a systematic approach to design, by Jones, Alexander, Archer, and Luckman, several common aspects become quite clear. For instance, there is considerable overlap in the reasons given for the emergence, and the necessity, of such systematic approaches. The late 1950s and early 1960s had seen increasing technological change and concomitant increasing complexity in the designer's task. Alexander refers to 'changes in technology and living habits happening faster all the time', and Luckman refers to a 'rapidly changing technological world'. There had also been the emergence of the systems approach to design, which is implicit in the procedures of all four authors, and

stated explicitly by Archer: 'The current tendency in design ...
is to try to consider the whole system of which the proposed
product is a part, instead of considering the product as a
self-contained object.'

The aims of systematic design procedures are, according to
Jones, 'To reduce the amount of design error, re-design and
delay', and 'To make possible more imaginative and advanced
designs.' And according to Archer, 'Systematic methods come
into their own under one or more of three conditions: when the
consequences of being wrong are grave; when the probability
of being wrong is high; and/or when the number of interacting
variables is so great that the break-even point of man-hour cost
versus machine-hour cost is passed.' There is, therefore, a
common concern with increasing both the efficiency and the
reliability of the design process in the face of the increasing
complexity of design tasks.

This common concern results in a considerable commonality
of approach. In particular, for all four authors there is an
emphasis firstly on extensive problem exploration and analysis
to identify all the factors that have to be taken into account, and
secondly on systematically establishing the interconnections
between all these factors so that all the sub-problems are
identified. They all also adopt the common approach of first
breaking down the overall problem into its sub-problems and
then attempting to synthesize a complete solution by combining partial solutions.

The fact that in four different fields—engineering design,
urban design, industrial design, and architectural design—such
similar approaches were being recommended lent support to
the notion that there is an underlying common design process.
However, the adoption of systematic design procedures was by
no means a rapid and universal event thoughout the several
design fields. What many designers—and design methodologists—failed to do was to heed the warning words of Jones:

The great difficulty of introducing Systematic Design is that its
advantages are not obtained unless it is carried out far more
thoroughly than is likely in first attempts. Successful application is
much more likely when changes in *organization* have been introduced
beforehand. As with many new things it involves an acclimatization
period during which things may get worse before they get better.

Further Reading

The papers by Jones and Alexander were presented at the first
conference on design methods, held in London in 1962. Many other
papers presented at the conference are still interesting and relevant; see
the proceedings, Jones, J. C., and Thornley, D. (eds) (1963),
Conference on Design Methods, Pergamon, Oxford.

Jones later made another significant contribution to design methodology with the first textbook of design methods: Jones, J. C. (1970),

Design Methods: seeds of human futures, Wiley, Chichester. For a more recent contribution by Jones to the management of the design process, see Jones, J. C. (1979), 'Designing designing', *Design Studies*, **1**, (1), 31–35.

The theory and background to Alexander's example of the design of an Indian village is presented more fully in Alexander, C. (1964), *Notes on the Synthesis of Form*, Harvard University Press, Cambridge, Mass. Some of the theoretical bases of this work (particularly its use of biological analogies) are criticized in Chapter 12 of Steadman, P. (1979), *The Evolution of Designs: biological analogy in architecture and the applied arts*, Cambridge University Press, Cambridge. An example of Alexander's method applied to American suburban house design is given in Chermayeff, S., and Alexander, C. (1966), *Community and Privacy*, Penguin, Harmondsworth.

A more detailed development of Archer's work is given in Archer, L. B. (1969), 'The structure of the design process' in Broadbent, G., and Ward, A. (eds), *Design Methods in Architecture*, Lund Humphries, London.

An early work on design process in engineering is Asimow, M. (1962), *Introduction to Design*, Prentice-Hall, Englewood Cliffs, New Jersey. For a methodologically orientated textbook in the architectural field, see Broadbent, G. (1973), *Design in Architecture*, Wiley, Chichester. A more general and more recent work is Nadler, G. (1981), *The Planning and Design Approach*, Wiley, New York.

1.1 A Method of Systematic Design

J. Christopher Jones

INTRODUCTION

A trend towards more logical and systematic methods of design has been evident throughout the 1950s. In many cases they have appeared as the result of new technical developments such as computers, automatic controls and systems. During the same period there have been attempts to give more scope for imagination and creative thought in design under such titles as 'creative engineering' and 'brainstorming'.

The following notes are an attempt to integrate all such developments as have come to the writer's notice into a unified system of design. It is aimed particularly at the area that lies between traditional methods, based on intuition and experience, on the one hand, and a rigorous mathematical or logical treatment, on the other. The method is intended to have two effects:

(1) to reduce the amount of design error, re-design and delay;
(2) to make possible more imaginative and advanced designs.

It is thought that the method can be usefully applied to design problems in any area in which:

(1) large quantities of design information are available, or can be usefully obtained;
(2) the design team has well-defined responsibilities, can

Originally published in Jones, J. C., and Thornley, D. (eds) (1963), *Conference on Design Methods*, Pergamon, Oxford. Reproduced by permission of Pergamon Press Ltd.

concentrate on development, and is free of routine design
work;

(3) considerable departures from existing designs are called
for.

Some parts of the suggested method (Sections 1.1, 1.2, 1.3,
1.4, 1.5, 2.2, 2.3, 2.4) have been tried out with some success in
design teaching projects and in prototype development in
ergonomics and industrial design. In these cases it has seldom
been possible to carry the method through to a logical
conclusion for want of the new kind of design organization that
seems to be necessary to permit a complete change to systematic work. Other parts of the method are based more on
conjecture, and on the reports of others, than on the writer's
experience. In many cases the suggested methods are better
regarded as a basis for further investigation than for application
in current design work.

Opportunities to try out some of the ideas presented here
have been provided by the Industrial Design Office at Associated Electrical Industries, Manchester, and by the Regional
College of Art, Manchester. For the patience with which those
concerned in both these organizations have awaited, and still
await, a substantial result from these activities, the writer is
very grateful. Perhaps this conference will mark the beginning
of a phase when systematic design methods will develop more
rapidly towards the degree of reliability and realism that is
necessary before they can be generally applied.

SUMMARY

The method is primarily a means of resolving a conflict that
exists between logical analysis and creative thought. The
difficulty is that the imagination does not work well unless it is
free to alternate between all aspects of the problem, in any
order, and at any time, whereas logical analysis breaks down if
there is the least departure from a systematic step-by-step
sequence. It follows that any design method must permit both
kinds of thought to proceed together if any progress is to be
made. Existing methods depend largely on keeping logic and
imagination, problem and solution, apart only by an effort of
will, and their failures can largely be ascribed to the difficulty of
keeping both these processes going separately in the mind of
one person. Systematic Design is primarily a means of keeping
logic and imagination separate by external rather than internal
means. The method is:

(1) To leave the mind free to produce ideas, solutions,
hunches, guesswork, at any time without being inhibited
by practical limitations and without confusing the processes of analysis.

(2) To provide a system of notation which records *every* item
of design information outside the memory, keeps design
requirements and solutions completely separate from each
other, and provides a systematic means of relating solutions
to requirements with the least possible compromise. This
means that while the mind moves from problem analysis to
solution-seeking whenever it feels the need, the recording
develops in three distinct stages:

1. *Analysis*: Listing of all design requirements and the
   reduction of these to a complete set of logically related
   performance specifications.
2. *Synthesis*: Finding possible solutions for each individual performance specification and building up complete designs from these with least possible compromise.
3. *Evaluation*: Evaluating the accuracy with which
   alternative designs fulfil performance requirements for
   operation, manufacture and sales *before* the final design
   is selected.

All the stages of SYSTEMATIC DESIGN are listed below. Details
of each stage appear on the succeeding pages.

1. Analysis	1.1 Random list of factors
	1.2 Classification of factors
	1.3 Sources of information
	1.4 Interactions between factors
	1.5 Performance specifications
	1.6 Obtaining agreement
2. Synthesis	2.1 Creative thinking
	2.2 Partial solutions
	2.3 Limits
	2.4 Combined solutions
	2.5 Solution plotting
3. Evaluation	3.1 Methods of evaluation
	3.2 Evaluation for operation
	for manufacture
	for sales

1. ANALYSIS

1.1. Random list of factors

At the first meeting each person writes a list of all the thoughts
that occur to him on acquaintance with the problem. Each
person reads out his list and all items are recorded in serial order
in a RANDOM LIST OF FACTORS. No attempt is made to avoid
duplication or to omit impractical ideas. In the ensuing
discussion there is a complete ban on criticism or comment on
the ideas produced. The object is to get down a large amount of

information in a short time in an atmosphere in which all feel assured that no idea will be inhibited or ridiculed.

The discussion can include, or be followed by, a similar random list of criticisms made while looking at examples, drawings and reports of the existing designs and at the ideas and information collected above.

Once the initial reactions and feelings about the problems have been recorded, the random list should be extended until it includes every single factor which could be thought to influence the design. This is to include all the obvious factors which may not have been mentioned in the initial discussion.

The notation for a RANDOM FACTORS LIST is to number each factor in the order in which they are first considered, e.g. some of the factors for the redesign of an electric motor might be numbered as follows:

(1) Improve output per unit volume.
(2) Access to terminals.
(3) Fit existing couplings.
(4) Overall height—reduce.
(5) Heat dissipation—method using less volume—faster coolant speeds.
(6) Existing design clumsy-looking.
(7) Why waste so much energy as heat—what would reduce the losses? List all possible items later.
(8) Shaft diameter, shaft length, keying method.
(9) Method of slinging.
(10) Can use of crane be avoided?
(11) Possible new materials, etc., up to a very large number of items, perhaps several hundred.

1.2. Classification of factors

At this stage, and at any later stages when the information may be too disorderly to reveal any general pattern, a CLASSIFICATION CHART can be used to find the alternative categories into which the factors fall (Figure 1).

The random factors are placed in categories by choosing a category title (category 1) for the first factor (factor 1), then choosing another category for factor 2, another for factor 3, etc., and denoting categories by crosses as shown. Factors may fall into categories already noted, e.g. factor 4 in category 2, or into more than one category, e.g. factor 6 in categories 1 and 3.

The factors are then copied out on separate sheets for each category (Figure 2) using the same factor numbers. It may be necessary to make another classification chart to find the subdivisions of categories containing many factors.

Usually the act of writing out the factors in separate categories suggests further additions and begins to show up areas where more information is required.

Factors	Categories					
	1	2	3	4	5	etc.
1.	x					
2.		x				
3.			x			
4.		x				
5.				x		
6.	x		x			
etc.						

Figure 1
Classification chart.

Category 1

1
6
15
etc.

Figure 2

One special category will be 'ideas, solutions, design propos-als, etc.'. These should be copied on to a separate sheet to which many more ideas will be added as the work proceeds. Ideas sheets should be kept completely separate from information and design requirement sheets—possibly on the opposite sides of a folder (Figure 3).

An alternative method is to write each factor on a separate index card and to sort these into alternative sets of categories and sub-categories until an acceptable arrangement is found.

1.3. Sources of information

The previous steps will have produced, from the existing knowledge of the design team, a very large quantity of

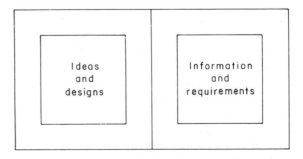

Ideas
and
designs

Information
and
requirements

Figure 3

information, much of it conflicting and in need of verification. Classification will have suggested further factors, will show up areas where further information is needed, and will also suggest likely sources, e.g. the category 'Maintenance' might show up ignorance of what the maintenance difficulties are and suggest consultation with, and observation of, maintenance staff. In any category the possible sources of information are literature, experienced persons, observations, and experiment.

Literature

Technical and trade journals, abstracts, research reports, technical libraries, etc., contain nearly all that is known about any existing problem. But the location of information relevant to one's own particular problem can be so time-consuming and expensive that it often seems more helpful to rely on one's own judgment or to make a new investigation oneself despite the risk of missing important technical developments and despite the additional cost of investigations which may already have been carried out, and paid for, by others. Some of the very real difficulties of using literature in design can be overcome in the following ways.

Use of libraries and information services. Technical libraries report that enquirers have difficulty in explaining their needs in a way which would make it possible for library staff to help them. Most people describe their interests too broadly, for fear that the librarian will miss something important, and so they get back a large quantity of material that is irrelevant or already familiar. The correct method is to tell the librarian not what you know but *what it is that you do not know* as precisely as possible, e.g. instead of asking for 'any information on the cooling of electric motors?' describe what it is you do not know in such terms as 'what are the costs and efficiencies of the various methods of heat transfer that have been applied to electric motors?'

The extraction of relevant information. The quantity of technical literature is so great that one cannot hope to remember or collect much useful information from general reading until one has a particular problem in mind, and by then there often seems to be no time available for such time-consuming browsing. One answer is to do a relatively small amount of intensive reading of a few publications while the random factors list is being compiled and note down all factors that are suggested by what is read. It is astonishing how much useful information, that one would normally have missed, comes to light when one has so definite a purpose in reading. This process can be usefully continued until factors begin to repeat themselves and new factors cease to be found very often.

Experienced persons

Vast quantities of valuable and highly relevant design information exists in the memories of persons experienced in design,
manufacture, operation, costing, sales, maintenance, installation, transport, etc. It is often not made use of because the facts
are so mixed up with personal opinions and because it is so
difficult to explain one's experience to someone else. However
it is worth trying the following methods of overcoming these
difficulties.

Separating opinion from fact. The obvious way to do this is to
ask each question of two or more people and treat as opinion all
those items on which they disagree. Items on which they agree
are then either facts, or opinions which are widely held. It may
be possible to sift out the latter by some objective checks and
measurements. When there is no means of checking, and when
experienced persons are the only available source, it is important to act only on the advice of persons who will afterwards be
held responsible for what they recommend. Opinions of any
other persons, whatever their status or authority, cannot be
incorporated into a design unless they are shared by a person
who is to be held responsible for their effect upon the design.

Judgment between limits. Experienced people are often reluctant to commit themselves to precise answers and instead give
very general answers that are of little help in design. The answer
here is to press for judgment between precise limits, e.g. an
experienced costing engineer may be reluctant to quote any
figure on the basis of a rough sketch but if asked to state the
limits of his uncertainty will be able to give a maximum and
minimum cost which will be a far more accurate guide than any
guess which the questioner could make. Almost any experience
can be expressed between limits if the question is properly
stated in terms of the quantity which will affect the design itself.

Logical question chains. Although opinions may be useless in
themselves they are often based on factual experience which can
be extracted by a chain of logically related questions, e.g.:

Opinion : Customers will not consider electronic controls
under any circumstances.
Question : Why will they not consider them?
Answer : Because they are unreliable.
Question : In what way are they unreliable?
Answer : They are always going out of order.
Question : But do not existing kinds of controls go out of
order?
Answer : Yes, but you have time to do something about it
before the thing packs up completely.

Question : Electronic controls give no warning of failure
then?
Answer : Yes—I suppose that is it—an electronic control
that gave a warning would be acceptable.

The logical question chain has tracked down the basic fact
which caused the opinion and which shows up a possible
method of overcoming what seemed at first like an irrational
opinion that prevented all progress in design.

Observation

The difficulty of observation is that in any practical situation
there is so much to see that one cannot select the facts that are
relevant to one's own problem. Some methods of overcoming
this difficulty are:

Recording for analysis later. The obvious way to overcome the
difficulty is to do the selecting of relevant facts at a later time
and in a less distracting environment. Two methods of
recording can be used.

(1) *Unabstracted recording* such as photographs, films and
general notes which record all that can be seen and leave a
great choice of facts from which to select for analysis. Such
methods can record significant information that was not
expected beforehand but involve lengthy and costly analysis and do not give very definite answers.
(2) *Abstracted recording* such as pen recorders, punched tape,
checking facts on prepared forms. Such methods omit
everything that was not expected beforehand but greatly
reduce the time and cost of analysis and give very definite
answers.

It is clear that initial observations should be of type 1 and that
later observations should be of type 2.

Experiment

It is sometimes necessary to resolve a doubtful point by direct
experiments or trials. As in the above cases of observation the
important things are to:

(1) state clearly what one does *not* know and within what
limits;
(2) calculate or estimate the limits of errors in the experimental
situation;
(3) before designing the experiment decide what actions will be
taken for each of the possible outcomes.

This procedure gives the greatest amount of useful information for the least expenditure of time and effort.

1.4. Interactions between factors

Many factors affecting design interact with one another to make up a complicated situation which can be appreciated by experience. The systematic approach is to use charts, to ensure that all possible interactions are discovered, and diagrams, to make clear the pattern of relationships.

There are many mathematical and diagrammatic methods of doing these things. Two very elementary methods which have very wide applications in the finding of relationships in randomly assembled information are given here.

All possible combinations, by pairs, of the factors 1, 2, 3 etc. appear in the squares to the right of the diagonal line of the interaction matrix (Figure 4). Each pair is considered separately and when an interaction is found it is denoted by a cross. Examples of the use of this chart are:

(1) Discovering all factors influencing an existing design by listing all the parts of the design, and all the elements with which it has contact, as factors. For every pair of factors which is found to interact there will be a requirement which can afterwards be expressed as a performance specification (see Section 1.5). This procedure is particularly useful when one or two design requirements have changed and it is necessary to find out how these changes will affect the rest of the design. Such a check will sometimes make it clear that the whole problem has also changed and an entirely new design is called for.

Factors	1	2	3	4	5	6	7	8	9	...
1				+				+		
2			+			+			+	
3							+		+	
4					+			+		
5									+	
6										
7									+	
8										
9										

Figure 4
Interaction matrix.

(2) When the design problem involves the linking of many
elements as in the case of circuits, pipework, building
layouts, communication systems. The required linkages
can be checked by listing all the parts of the system as factors
and deciding whether or not a link is required on every
possible interaction.

Interaction matrices can also be used to establish sequences and
orders of importance factors. In this case the whole area of the
chart is used (Figure 5).

Factors	1	2	3	4	5	6		
1 C		+	+			+		
2 F								
3 D		+				+		
4 A	+	+	+		+	+		
5 B	+	+	+			+		
6 E		+						
Totals	2	5	3	0	1	4		

Figure 5

   Each cross denotes that the factor in the rows across comes
before the corresponding factor in the column, e.g. the cross at
square 1–2 means that factor 1 comes before (or is more
important than, or has some kind of sequential effect on) factor
2. The totals give the final sequence of factors 1–6 namely:

Total	0	1	2	3	4	5
Factor	A	B	C	D	E	F

This method of establishing sequences can be applied to such
things as:

(1) Finding the operating sequence for a number of controls.
In this case each cross indicates that the control in the row
affects the operation of the control in the column but not
vice-versa.
(2) Placing a number of design requirements in order of
importance. This is a good method for weighting factors
prior to decision making.

Interaction nets

The pattern of relationships between a number of interacting
factors can be made easy to see by replotting the information
from a matrix as a net (Figure 6).

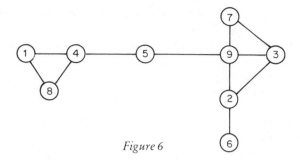

Figure 6

Each number refers to a factor and each line refers to a cross in the interaction matrix. The above interaction net is obtained in two stages.

(1) Write out the numbers in a circle and draw in the links, corresponding to the crosses, with as few crossed lines as possible—starting with numbers which have many links (Figure 7).

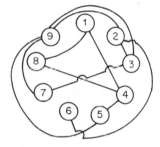

Figure 7

(2) Re-draw in an orderly way keeping groups of interconnected numbers apart from each other and with as few crossed lines as possible, as in Figure 6. In very complicated cases confusion can only be avoided by the use of a three-dimensional model.

Examples of the use of interaction nets are:

(1) Making clear the pattern of interactions between all the performance specifications in a problem and thus giving an idea of the consequence of changing any particular requirement.

(2) Making clear the pattern of linkages in any problem that involves the interconnection of many parts, e.g. circuits, systems, circulation plans.

1.5. Performance specifications (P-SPECS)

Complete separation of problem from solution can only be achieved if the requirements are expressed purely in terms of performance and with no reference whatever to shape, materials and design. To do this each requirement is rewritten as a PERFORMANCE SPECIFICATION (or P-SPEC), e.g.:

> Design specification:
>   Control panel to be mounted at 45°
> P-SPEC:
>   Control panel to be visible from all operating positions.

It is often difficult for those who are accustomed to working by experience rather than analysis to think in terms of performance by itself. This difficulty can often be overcome by subjecting each requirement to a LOGICAL QUESTION CHAIN (see Section 1.3).

P-SPECS should be expressed wherever possible between LIMITS (as in Section 1.3), e.g.:

> P-SPECS between limits:
>   Control panels to be visible within ±20° of the normal from all positions of the eyes for persons of height between 5 ft 3 in. and 6 ft 2 in.
>   Heat dissipation to be between 10 and 20 per cent more than for existing design.

In all cases the limits should be as far *apart* as is in keeping with the standards of performance aimed at.

1.6.   Obtaining agreement

The complete list of P-SPECS is circulated to all persons whose agreement to the final design will be required at a later stage. At subsequent meetings and discussions every effort is made to incorporate all valid criticisms and omissions in a revised list. Written agreement of all concerned is sought at this point on the strict understanding that the design will go ahead on this basis and all further alterations or additions will have to be held over for incorporation in future designs.

The strict finalizing of the P-SPEC list will only be possible when the preliminary work has been very thorough. When this degree of thoroughness has not been achieved, major errors and omissions in the P-SPECS are likely to be discovered at later stages. In these cases it will not be possible to maintain separation of problem and solution and SYSTEMATIC DESIGN will have ceased to occur. This failure is likely to occur when SYSTEMATIC DESIGN is tried out for the first time in situations where traditional methods are customary. The great difficulty of introducing SYSTEMATIC DESIGN is that its advantages are not obtained unless it is carried out far more thoroughly than is likely in first attempts. Successful application is much more likely when changes in ORGANIZATION have been introduced beforehand. As with many new things it involves an acclimatization period during which things may get worse before they get better.

2.   SYNTHESIS

2.1.   Creative thinking

Little is known of what goes on in the mind during creative thinking, but it is clear that the power of imagination which any person can apply to the solving of a design problem is dependent on two external factors:

(1) A clear statement of the problem.
(2) A free atmosphere in which any idea, however crude or naïve, can be expressed at any time without regard to its practicability. The purpose of all the previous stages is to make possible a very clear statement of the problem. This work will in itself suggest many ideas and solutions which will have been collected in a separate set of papers as described in Section 1.2. When the P-SPECS have been completed a much more deliberate search for solutions can begin.

A useful technique is that described as BRAINSTORMING. Those concerned with the project, and a number of other persons of different backgrounds and experience, are invited to an informal meeting. After each item of the performance specification is read out all persons are asked to mention any idea, criticism, or solution that comes into their heads, however mad or eccentric it may appear to be. The chairman does nothing except read out the problems and enforce the rule that nobody shall criticize any idea in any way. All ideas are recorded in shorthand or by tape recorder for evaluation later.

The same method of uninhibited listing of ideas can be done by individuals. This procedure is valuable as a means of releasing the imagination and of obtaining a large number of ideas from many areas of experience in a short time. It should be followed by more logical and careful thought, particularly on those aspects of the problem where no obvious solutions have appeared so far. In such cases, if the imagination comes to a halt and no solution seems possible the following methods may suggest a solution:

(1) Write down the conditions which would make a solution possible.
(2) Write down a phrase describing the difficulty and substitute alternatives for each word, e.g.:

PHRASE DESCRIBING A DIFFICULTY

'Manufacturing tolerances will cause misalignment between two surfaces.'

ALTERNATIVE WORDS

Substitute 'welding distortion' for 'manufacturing toler-ances':

This suggests using a process other than welding.

Substitute 'a gap between' for 'misalignment':

This suggests making a deliberate gap.

Substitute 'adjacent surface' for 'two surfaces':

This suggests making the surfaces not adjacent, i.e. putting another surface between them.

Substitute 'planes' for 'surfaces':

This suggests making the surfaces non-planar.

(3) Write down the consequences of not finding a solution of the difficulty. This will often show that the consequences can in fact be tolerated, or suggest another way of avoiding them.

2.2. Partial solutions

Existing design methods usually result in a single solution which is conceived as a whole with the details being worked out later. SYSTEMATIC DESIGN reverses this procedure. PARTIAL SOLUTIONS, one or more for each P-SPEC, are sought first and then combined by permutation to give several alternative whole solutions.

Partial solutions to each of the P-SPECS are tabulated as shown in Figure 8.

	P - SPECS	Partial Solutions
1		
2		
3		

Figure 8

A vital difference between this and existing methods is that each PARTIAL SOLUTION is considered *completely independently* of any other. The object is to find shapes which meet each P-SPEC in the best possible way and to take no account of conflicting demands of different P-SPECS, e.g. if one P-SPEC is best satisfied by a removable access cover and another is best satisfied by a fixed enclosure these best solutions should be separately recorded, together with some estimate of the degree of performance loss for departures from complete accessibility and complete enclosure respectively. This makes it possible to find COMBINED SOLUTIONS having the *least compromise* at stage 2.4.

Sources of partial solutions are:
(1) A search through the ideas collected so far.
(2) The use of BRAINSTORMING, and the other more logical methods described in Section 2.1 as aids to creative thinking.
(3) Observation and experiments with existing designs and experimental apparatus to measure the degree to which any particular P-SPEC is satisfied, without reference to any other.

All solutions should be expressed in terms of physical dimensions between limits as described in the next section.

2.3. Limits

There will be a *range* of dimensions, shapes, and variations in material properties, which will satisfy any particular P-SPEC. It is vital to express PARTIAL SOLUTIONS by LIMITS which show clearly the extremes between which all such equally acceptable solutions lie. This means that:

(1) All dimensions, radii, etc., to be followed by the widest acceptable tolerance, e.g.:
$+ 1$ in. $- 1$ in., $+ 1$ in. $- 0$ in., $+ 0$ in. $- 1$ in.
(2) All shapes to be expressed as a range of shapes, e.g. any shape lying between a square and a circle.
(3) Choice of materials to be expressed between limits of their properites, e.g. any materials have $A \pm a$ hardness, $B \pm b$ elasticity, etc.

2.4. Combined solutions

The purpose of this stage is to combine PARTIAL SOLUTIONS into COMBINED SOLUTIONS with the least departure from the P-SPECS. This can be done to some extent by looking down the list of PARTIAL SOLUTIONS and selecting by eye those that appear to go best together. But as this process involves far more alternatives than the memory can handle without confusion it is necessary to do this in stages.

(1) By inspection, and a little trial and error, make lists of
(a) PARTIAL SOLUTIONS which are compatible, i.e. do *not* conflict.
(b) PARTIAL SOLUTIONS which are incompatible, i.e. *do* conflict.
An INTERACTION MATRIX (Section 1.4) can be used to check the compatibility of each pair of solutions.
(2) Look again at the INTERACTIONS BETWEEN FACTORS (Section 1.4) to find logical sequences in which PARTIAL SOLUTIONS should be combined. For example, in the design

of an automatic control system the INTERACTION NET might show that:

(1) The *type of switching element* is dependent on:
(2) The *frequency of switching,* which is dependent on:
(3) The *speed of response,* which is dependent on:
(4) The *acceptable error,* which is dependent on:
(5) The *relative costs of errors* and *error-correcting devices*:

It is correct to begin with the most independent factor, in this case the cost of errors, and to combine PARTIAL SOLUTIONS in the sequence 4, 3, 2, 1, 5, repeating the process until the optimum for 5 is obtained. In this way sets of compatible particular solutions, each following within the LIMITS of their respective PARTIAL SOLUTIONS, can be obtained.

Compatible PARTIAL SOLUTIONS can be tabulated as in Figure 9.

P - SPECS	Sets of compatible partial solutions					
	Set 1	Set 2	Set 3	Set 4	Set 5	
1	P11	P12	P13	P14		
2	P21	P22				
3	P31		P33			
4		P42	P43	P45		

Figure 9

P12 is a partial solution. The figure 1 indicates that it falls between the dimensional limits of the PARTIAL SOLUTIONS for P-SPEC No. 1. The figure 2 indicates that it is compatible with all the other solutions in set 2, compatibility having been discovered by both or either of the methods 1 and 2 above. It may be more convenient to put small illustrations on the chart in place of the symbols P12, etc., and so make it possible to see at a glance the *way* in which the particular solutions will fit in with each other.

As the gaps in the above chart indicate, no set of compatible solutions will meet all the P-SPECS. The problem then is to decide which combinations give least compromise and loss of performance. Often the only way to do this is to develop several sets to the stage where their performance can be evaluated by trial and experiment.

The design method described so far will still leave a great many detailed decisions to be made by eye and by experience.

2.5. Solution plotting

Systematic design does not seek a single solution but aims at finding a range of solutions and making it clear in what ways each fits, or does not fit, the specification. Finally, when it becomes necessary to select one or more of these solutions for production, the selection can be made in full awareness of the results which can be expected, and with a high probability that no better solutions can be found with the existing resources.

Solution plotting is a means of making clear the relationships between solutions. Two kinds of relationship are considered here.

Trends—i.e. relationships between previous solutions.

New solutions—i.e. relationships between all possible alternative solutions.

A trend plot can be used to show how shape and performance have changed over the years (Figure 10). Photographs or drawings of previous designs are placed in sequence and data are added in terms of reasons for design changes and details of performance achieved.

Year	Year 1	Year 2	Year 3	Year 4	Etc.
Illustrations	Original design	—	First design change	Second design change	—
Reasons for design changes	—	—	Reasons	Reasons	—
Performance for Factor 1	Performance	—	Performance	Performance	—
Performance for Factor 2	Performance	—	Performance	Performance	—
Etc.					

Figure 10
Trend plot.

The gaps which occur in years where no design change takes place serve to illustrate the life of each design and are helpful in indicating the degree of design change that is appropriate at the present time.

A new solution plot is a means of comparing the range of existing solutions, or new proposals, in relation to shape or performance (Figure 11). These plots can be used to find areas where new combinations of shape and performance can be sought and will also show up areas where the existing solutions are likely to be the best. New solution plots for both shape and performance take the form of graphs in which measures of two major characteristics of shape or performance are marked along the axes. Each of the points S1, S2, etc., marks the characteristics C1 and C2 of solutions. C1 and C2 can be of three kinds:

(1) two measures of *performance*, e.g. speed and reliability, cost and capacity;
(2) two measures of *shape*, e.g. weight and length, volume and type of material;
(3) a measure of *performance* and a measure of *shape*, e.g. speed and weight, maintainability and degree of enclosure.

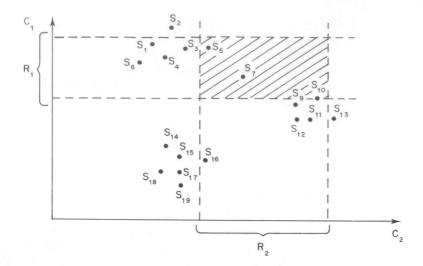

Figure 11
New solution plot.

The measures used to quantify characteristics C1 and C2 are either existing scales, such as weight and cost, or ranks obtained from making estimates and judgments between every pair of solutions using an interaction matrix (Section 1.4) to rank them in sequence.

The results of plotting all the alternative existing solutions S1, S2, S3, etc., might be patterns such as that shown. In the above example one might conclude: that existing shapes and materials tend to three basic design types, represented by the three clusters; that the central designs in each cluster may be the best obtainable example of their types; that isolated examples, such as S7, may be hybrids which will move towards one of the neighbouring clusters if modified to increase either of the characteristics C1, C2.

When the ranges of desired characteristics, shown as R1, R2, are known, the area in which both occur can be marked on the solution plot. In this case the position of the shaded area shows that only the fringe solutions S5, S10 and the hybrid solution S7 fall within it. This might suggest a search for a possible new cluster of designs towards the top right corner of the shaded area.

3.   EVALUATION

3.1.   Methods of evaluation

Evaluation is here understood to mean any method by which deficiencies in solutions can be detected *before* final manufacturing drawings have been started, *before* production begins, *before* the product has been sold, *before* it has been installed and *before* it has been put to use. The purpose of any method of evaluation is to detect errors at the stage when they can be most cheaply corrected, that is when the increasingly expensive processes of drawing, manufacture, selling, installation, and use do not have to be repeated in order to make the correction.

The traditional method of evaluation of engineering designs is by judgment, and by reference to the experience of engineers and draughtsmen, while the design is on the drawing board. When the right kind of experience is available, and when logical methods of detecting errors are too expensive or time-consuming, this is still the most effective method. The traditional engineering drawing can thus be regarded as a complicated two-dimensional model which permits the problems of manufacture, selling, installation, operation and use to be anticipated, and the design evaluated, at an early stage.

We are, however, already approaching a situation where engineering is called upon to design and develop increasingly complex equipment of which little or no experience exists and for which engineering drawings do not provide an adequate means of evaluation. At the same time we are becoming accustomed to using more expensive, more logical and more exhaustive methods of evaluation such as field trials, market surveys, models, simulators, computers, operational research, pre-production, pre-engineering and product planning. The purpose of this and the following sections is to summarize what is generally known of these new methods and to show how they relate to the techniques described in these notes.

Evaluation by P-SPEC

The list of performance specifications (Section 1.5) should include a large proportion of the items to be tested by evaluation. The major task then consists of devising tests, trials, or other methods by which each design is checked against each performance requirement. It is helpful if the P-SPECs have been worded with such tests in mind, e.g. 'accessibility *such that* terminals can be checked in less than 1 min'; 'strength *such that* part A deflects no more than 0.01 in. relative to part B when maximum load is suddenly applied'. Section 1.3 is relevant here.

Evaluation by judgment

In engineering design as it is today with new and unfamiliar logical methods being used here and there, with an increasing demand for the integration of previously separate components into complex interacting systems, with the continued raising of standards of performance and efficiency, and with the general lack of time, development funds, and information which would make completely logical decisions possible, it is necessary to use a mixture of experienced judgment and logical decisions in the evaluaton of any one design. These notes can be regarded as an attempt to bridge the very considerable gulf between these two methods and to overcome the inherent difficulties of using experience and logic side by side in one design problem. Some suggestions for mixing experience and logic in evaluation are given below.

Making judgments more precise. Stating judgments between limits; stating judments in negative form (i.e. stating what must be excluded); putting judgments in order of importance using paired comparisons in an interaction matrix; separating opinions from facts based on experience by consulting several persons, or by logical question chains. All these methods have been referred to previously. Another is to list the consequences of *disregarding* each judgment.

Extending the area of judgment. The main method here is to bring the experience of many persons to each valuation that has to be made. This can be done by a design review committee of senior and experienced persons and perhaps including a few persons of experience in non-engineering fields such as scientific research, personnel management, accounting, safety, etc. Such a committee would have no responsibility for design decisions. Its responsibility would be to list the possible consequences of the solutions and decisions submitted to it for judgment and to predict the sub-problems to which any solution or decision may give rise.

Separation of logical decisions and judgments. When a largely systematic design method is being used it may well be helpful to organize a new 'systematic design section', composed largely of young persons, responsible for using the method and making all the design decisions, but with the overriding rule that they must be able to demonstrate that all available sources of experience and judgment have been consulted—as in the design review procedure described above.

3.2.   Evaluation for operation, manufacture, and sales

The whole of science and technology can be regarded as a search for exact means of predicting the behaviour of physical things. For this reason all the methods of calculation and experiment of which engineering training is composed are relevant to the evaluation of designs and obviously include far more than can be summarized in these notes.

There is, however, one major aspect of calculation and experiment which is largely absent from current engineering training and which is of central importance to a systematic design method. That aspect is the prediction of the behaviour of *large numbers* of related components, over *long periods of time*. Traditional engineering calculations are largely confined to the prediction of isolated events and separate parts of a design problem and leave to human judgment the assessment of the behaviour of the design as a whole. This may be the reason why so much engineering can at present be done with hardly any calculation at all.

We are, however, coming nearer to a situation in which evaluation and prediction of the performance of designs as a whole, and for their complete life spans, is falling outside the range of experience. Newer and more comprehensive methods of calculation are coming into being, largely because of the availability of high-speed computers. But before these sophisticated methods can be introduced into more intangible and complex areas of judgment it is both necessary and possible to make a start with simpler and more 'manual' methods of logical evaluation, some of which are outlined below.

This 'manual' or semi-logical approach to evaluation could be the responsibility of the separate 'systematic design section' referred to in Section 3.1. Its basic methods of evaluation would be:

(1) Collection and assessment of *available experience and judgment* (using the methods of Section 3.1).
(2) *Simulation* (using any form of model making, drawing, analogue computing and experiment that is available and appropriate).
(3) *Logical prediction* of all possible circumstances and situations which the design has to meet through its life (using interaction matrices and nets to check all the permutations).
(4) *Pre-engineering development* (using small-scale but realistic facilities for pre-production, pre-sales and pre-operation in which unforeseen difficulties are discovered before full-scale production, sales and operation has started).

Item 4 entails the provision of fairly elaborate facilities of a kind not usually available to engineering development sections, as they are organized at the moment. These facilities are listed below.

Pre-production. The facility required is a small 'production development section', separate from the main shops, but operating to the same standards of accuracy (or inaccuracy) and under analogous limitations. Production prototypes and parts should be made without any of the additional hand work or fitting that is usually expected in prototype manufacture. This section to be under a staff of engineers trained in design, method study, and production engineering.

Pre-sales. The facility required is a small 'sales development section', separate from current sales and publicity work, which is in close touch with those sections of customer organizations that are most interested in future rather than present developments, and has the responsibility for getting pre-engineered prototypes accepted for trial by such customer groups. The section would make use of this situation to discover all the favourable and unfavourable sales reactions at this stage and would also prepare development versions of all the necessary sales literature, leaflets, advertisements, etc. This section would be the one most concerned with the making of realistic appearance models and with industrial design. It would be staffed by sales engineers with experience of systematic marketing methods and with some experience of design.

Pre-operation. The facility required is a small 'operation development section', separate from current engineering. It would work in close contact with planning and research sections of customer organizations. It would seek facilities for the small-scale trial of pre-engineered prototypes for the expected range of operation conditions these facilities being provided by customers themselves. The section could be staffed by engineers with experience of design, systems engineering and research. It would be the section most concerned with research and with ergonomics.

REFERENCES

Many of the methods described in the above notes have appeared in various publications during the past ten years and particularly in the American journals *Product Engineering* and *Machine Design*. Some of these are listed below:

Du Bois, G. B. (1953), 'Instruction in creative mechanical design', *ASME*, Paper No. 53–A–172.

Froshaug, A. (1958), *Visual Methodology*, ULM 4, Hochschule für Gestaltung, Ulm, W. Germany.

Jones, J. C. (1959), 'A systematic design method', *Design*, April.

Jones, J.C. (1961), 'The design process', Internal Report of AEI Industrial Design Conference.

Kay, R. M. (1959), 'Steps in designing', Industrial Design Office, AEI, Manchester.

Marples, D. L. (1960), *The Decisions of Engineering Design*, Institution of Engineering Designers.

Marvin, P. (1957), 'Engineering the creative process' *Machine Design*, 10 January.

Marvin, P. (1956), 'Picking profitable products', *Machine Design*, 9 Feb.

Murdick, R. G. (1961), 'The three phases of managerial planning', *Machine Design*, 13 April.

Nelles, M. (1953), 'Deliberate creativeness in engineering', *ASME*, Paper No. 53–A–171.

Osborn, A. F. (1963), *Applied Imagination—Principles and Procedures of Creative Thinking*, Scribener's Sons, New York.

Quick, G. G. (1961), 'Predict product performance', *Product Engineering*, 24 April.

Randsepp, E. (1959), 'The creative engineer', *Machine Design*, 28 May, 11 June, 25 June.

Shapero, A. and Bates, C. (1959), 'A method of performing human engineering analysis of weapon systems', WADC Technical Report 59–784. Office of Technical Services, US Dept of Commerce, Washington, DC.

1.2 The Determination of Components for an Indian Village

Christopher Alexander

INTRODUCTION

I. I shall discuss the general problem of finding the right physical components of a physical structure (or alternatively, the physical subsystems of a physical system).

II. This problem becomes especially urgent when we are faced with the development of a dynamic structure like a city; as I shall try to show, the nature of its components play a critical part in determining the rate and efficiency of its adaptation to new developments and changing situations.

III. The method I shall describe is based on non-numerical mathematics. It arises largely from the use of graphs (topological 1-complexes) to represent systems of interacting functions. Since the application of the method demands rather a lot of computation and manipulation, I have used the IBM 7090 computer to carry it out.

IV. For the sake of example I have chosen to work with a city in miniature, rather than a full-fledged city. During the last six months I have been working in an Indian village with a population of 600, trying to determine what its physical components should be, if its development and reactions to the future are to be efficient. I have some diagrams illustrating what I have done so far. In about a year's time I shall be going back to India to rebuild the village according to these principles.

Originally published in Jones, J. C., and Thornley, D. (eds) (1963), *Conference on Design Methods*, Pergamon, Oxford. Reproduced by permission of Pergamon Press Ltd.

I. COMPONENTS

We sometimes forget how deeply the nature of an object is determined by the nature of its components. Once you decide that a car is to be made of four wheels, engine, chassis, and superstructure, there are really very few essential changes you can make. You can alter the shape of the components, and the way they are put together; but what you have remains a car, as we have known it now for fifty years, even though it may be this year's model rather than last year's. Or take an example from elementary chemistry. The difference between graphite and diamond is very largely caused by the difference in the structure of their component building blocks. Both are pure carbon; but in the case of diamond, where the carbon atoms are arranged to form tetrahedral components, the ensuing structure is hard and brilliant, while in the case of graphite, where the atoms are arranged to form flat plate-like components, the resulting substance is soft and black and slippery.

Now think of an example from architecture. If you ask anyone to name the pieces a city is made of, he will tell you that it is made of houses, streets, factories, offices and parks, and so on. This very obvious fact was formally recognized in the Athens charter of CIAM in 1929, which stated that the four functions of a city were work, dwelling, recreation, and transportation, and that the physical articulation of the city (that is, its breakdown into components) should follow this division of functions.

When we ask what it is that makes each one of these macro-components what it is, we see that, just as the component plates of graphite are themselves made up of carbon atoms, so each component of the city is really a collection of smaller elements, and that each gets its character from the way these smaller elements are grouped. A house is a collection of bricks and heaters and doors and so on, grouped in such a way, and with the divisions between groups, and the relations between different groups arranged in such a way that we call the thing a house. If the bricks are arranged differently we get a wall; if less homogeneously, we get a pile of rubble, or a path. It is not hard to see that these bricks and other pieces can be put together in millions of different ways, and that the macro-components might therefore be very different from those we are used to. There are, very likely, other, newer, macro-components, latent in the life of the modern city, which would serve better as the city's building blocks than the old ones do; but their growth is hindered by the persistance of the old roads and factories and dwellings, both on the ground, and in the designer's mind. The habit of mind which says that the city's component building blocks are houses, streets, parks and offices, is so strong that we take the components for granted: what we call design, consists of playing variations on the kinds of city you can get by arranging and rearranging these components. But this never leads to a new structure for the city. Whether the streets are curved or straight, or long or short, the houses large or small, the park in the middle of the city, or on the edge, the behaviour of the whole is fundamentally unchanged.

Not only are planners and designers often trapped by the persistance of the known components, but those systematic techniques under discussion at this conference often aggravate their fault. Statistical decision theory applied to transportation may tell you to put a road here rather than there. Linear programming applied to the distribution of apartments, family dwellings, or an office block, might give you an optimum distribution for them. Location theory might tell you where to put a park so that it will not inhibit the economic growth of a city, and yet serve the optimum number of people. Other theories will tell you how many floors an office block should have, how large to make the car parks, how many lanes to give the superhighways, and so on. But all these methods, though they are perhaps more precise than an average designer's intuition, still leave the essential structure of the city the same as it was before. If you once agree that houses, streets, parks, and offices, are the proper components of the city (as you have already done, usually, when you even begin to use some systematic method), then there is really very little choice open to you about the city structure. Systematic techniques, just because they need to operate on known units, usually beg the real question of design, and so achieve little more than a second rate designer does. The fundamental change which a structure undergoes at the hands of a great designer, who is able to redistribute its functions altogether, cannot take place if its components stay the same.

II.   THE CITY AS AN ADAPTING SYSTEM

Let us now discuss the special nature of the city problem. A city is a live structure, not a dead one. That is, it is a loose assemblage or aggregate of components, which is all the time being added to, and changed.

The form of a city cannot bear the same relation to its needs, as a designed object does. Since the life of the city is always much greater than the life-span of any one pattern of needs, the needs of the city change constantly and unpredictably through-out its lifetime. It is entirely inadequate to treat the city as a building, and to design it by meeting any single pattern of needs, as was done in the case of Brasilia or Chandighar. Even the principle of the master plan, as it is usually practised, which takes into account projected needs, is only a little better, and contains the same fundamental mistake. The span of possible projections into the future is also limited—say it is thirty years.

Even a master plan which looks as far forward as thirty years from now, is still essentially static in its conception; it is still based on the fixed list of needs which are felt now and expected during the next thirty-year interval; whether the programme of needs to be met covers a few years more or less, cannot alter the fact that it invites a design conceived only to meet those fixed needs. If we are honest we know that we cannot see into the future properly; we know that in the future, this city which we are designing now, will generate needs and stresses unimaginable in the year of its design, not calculable by projections into the future; with changes in technology and living habits happening faster all the time, we must devote our attention to the development of what we might call a self-organizing city structure which, apart from meeting present needs, is by nature able to adapt easily to any changes in the pattern of needs whatever, foreseen or unforeseen.

I repeat what I said at the beginning. The city is a live assembly or aggregate of components. The problem of city design is not to design the city as a whole, but really to establish a kit of components out of which such a growing, ever-changing aggregate can be built up. These components will be determined by two properties which the aggregate system must have if it is to maintain its functional efficiency while expanding and changing to meet new needs and circumstances. The first property concerns the addition of new components to the system. The second concerns the modification and replacement of components already in the system.

I will deal first with the addition of new components. It is very certain that we cannot develop all the components of the city simultaneously. We do not have the money, nor the power, nor would it be desirable to cause such an upheaval in the city's life, nor does this correspond to the realities of a city's growth. The city develops gradually. To safeguard its development (as against design), we must define addable 'units' which are sufficiently self-contained, to be used as units of development and aggregation.

We may assure ourselves of the need for this, by looking at some examples from common experience. You may be able to put an artifical heart in a human body. But it would be almost impossible to replace three-quarters of the heart by an artificial three-quarter heart, and leave the other quarter of the natural heart functioning properly together with it. If you enlarge a golf course, you can add an integral number of holes; but you can hardly add a hole and a half. If you prune a tree, you make sure that you cut off limbs and branches at right angles to their length; you know that if you remove half a branch, split lengthwise, you have not left a properly functioning unit. If a foundry and a machine-shop are so well tied together that the machine shop can exactly cope with the castings which the foundry makes, then it is useless to add lathes to the machine shop, without also enlarging the foundry so as to increase its output. Although a lathe looks like a proper 'unit' of development, it is not a proper functioning unit of the 'machine-shop–foundry' ensemble.

This leads to the first axiom: *We cannot add unregulated quantities of elements to a developing structure, but only certain well-integrated units which bear a proper relation to the presently functioning whole. And we cannot change or replace arbitrarily chosen pieces of a functioning whole, but again, only units which are sufficiently well-integrated to function as units or components of the whole.*

It is easy to see that the components defined by work, dwelling, recreation, transportation, are not properly functioning units in this sense. When 50 million families have television in their homes, recreation and dwelling can hardly be treated as separable components. Again, the growing popularity of industrial ring roads like those round Boston or London, makes it doubtful whether work and transportation can be developed as separate components.

The second kind of development which shapes the city, is the modification of the existing fabric as the system encounters new internal needs, and tries to adapt them. It may be easiest to focus on this problem, by seeing what happens when it goes wrong.

Suppose a new kind of centrally distributed piped heat for dwellings becomes commercially feasible. With the city organized as it is today, it would be impossible to install this new piped heat in existing neighbourhoods, because of all the roads and gardens which would have to be dug up. In other words, the installation of piped heat calls for modification of so many of the existing components (dwellings, gardens, streets), that it simply cannot take place.

Take another example. Suppose a new kind of passenger vehicle is invented. It could be very hard for the existing city to adapt to it, because its garaging would affect the dwellings, its circulation would affect the streets, and its storage away from home would affect the recreation and work components. To provide for the new kind of vehicle, so many changes would have to be made, in so many parts of the city, that the city would come to a standstill while these changes were effected.

The same thing will happen when the working week is shortened, and puts more spare time on people's hands. It will have repercussions throughout the city: recreation facilities, homes, roads, will all be affected; again with the result that the modifications will not be made, because they would be too complicated and too widespread to carry out.

In all these cases, instead of the city adapting to the changing way of life, it is the people who bear the brunt of the adaptation by having to carry out new activities in obsolete surroundings. The city itself is unable to adapt to new circumstances as fast as they occur, because each one demands changes in nine-tenths of the city; such extensive changes are impossible, and therefore do not get done.

This failure of adaptation on the part of the city is not caused by the faulty design of the components. You cannot improve it by building better roads or better houses or better parks. The real trouble lies with the root choice of the components; the functions they represent, though perhaps independent to some extent, are not independent enough. What we require, ideally, are components which are so independent, functionally, that every new need and new situation which occurs, only demands change or modification in one of the existing components, instead of all of them. For the sake of example, let us go back to the piped heat. Suppose it were generally recognized, as it should be, that the distribution of telephone, water, mail, and other things from central sources to individual houses, constitutes a bundle of functions, strongly interdependent, and independent of other functions, so that some kind of huge ducting component was a normal part of the city and took care of all these things simultaneously. When the possibility of piped heat came up, it would then be necessary to modify only this one component. All the necessary changes would concern the duct, and nothing else; and as a result the change could be effected; and the adaption to the new situation could actually take place.

The general statement of this principle, which will allow the city to adapt rapidly to future changes, is my second axiom. *If a changing system in contact with a changing environment is to maintain its adaptation to that environment, it must have the property that every one of its subsystems with an independent function is also given so much physical independence as an isolable component, that the inertia of those components which for the time being require no modification, does not make it impossible to modify those other components which do need to be changed.*

To achieve the two properties required by adaptation, this one which I have just stated, and the other stated earlier, the components of the city must themselves have two properties. To satisfy the first axiom, the components must be functionally compact—that is, each must operate as a unit. And to satisfy the second axiom, the components must be highly independent of one another, functionally.

III.   THE METHOD

I shall now describe a method of determining components to meet these two conditions. It rests on the assumption that each component of a physical structure represents what we might call a 'coming together' of a number of functional requirements.

A street is the coming together of the need for circulation space, the need for access to houses, the need for light and air between buildings, the need for somewhere to bury services, the need to lead off excess water during rainstorms. All these requirements, because they have similar physical implications, together define the linear component of the city which we call a street. A house is the coming together of the need for somewhere to sleep, the need for a place to store and prepare the food, the need for a place to belong to, the need for a place to raise children, the need for protection from wind and rain. All these requirements come together to form the component of the city we call a house.

A less homely way of putting this is to say that each component of the physical city comes into being to deal with some specific subset or subsystem of requirements in its environment. If we want the components to have the kind of independence and unity described above, the way to ensure this is to put them in 1–1 correspondence with the most independent subsystems of their environment. In other words, establish the truly independent subsystems of requirements in the environment, and then match each one of these subsystems with a physical component of the physical system you are designing. I do this as follows.

Establish, first of all, the requirements which have any bearing on the physical shape of the village. The list which follows includes requirements of every conceivable kind: (1) all those which are explicitly felt by villagers themselves as needs, (2) all those which are called for by national and regional economy and social purpose, and (3) all those already satisfied implicitly in the present village (which are required, though not felt as needs by anybody). Each requirement must be clearly enough defined for it to be decidable in principle in an actual village, whether the requirement is satisfied or not. Beyond that the requirements need have nothing in common. In particular, it is not necessary for each requirement to have a quantifiable performance standard associated with it. In other words, all the variety and vagueness of requirements which we encounter in real world problems can be included.

[In the original there follows a list of 141 requirements.]

Secondly, in order to examine the system properties of the set of requirements, we must decide, for each pair of requirements, whether they are dependent or not. Again, the interactions I accept are not limited to those which can be expressed in the form of equations or other kinds of mathematical functions. Two requirements are dependent if whatever you do about meeting one makes it either harder or easier to meet the other, and if it is in the *nature* of the two requirements that they should be so connected, and not accidental.

Here is an example. 94 is the need for provision for animal traffic. This conflicts with 7, the need for cattle to be treated as sacred, because the sacredness of cattle allows the cattle great freedom, and hence more room for circulation, which makes 94 harder to meet adequately. On the other hand, 94 connects, positively, with 13, the need for family solidarity, because this latter requirement tends to group the houses of family members in compounds, and so reduces the number of access points required by cattle, making 94 easier to meet.

[In the original there follows a complete list of interactions.]

If we think of the requirements as points, and of the dependences between requirements as links, then the set of requirements and the set of links together define a linear graph (or topological 1-complex). This serves as a complete structural description of the functional environment which contains the village and calls it into being.

The beauty of this description is that we can now give it a mathematical interpretation, compatible with the real-world facts, though nonetheless artificial, which suggests criteria for decomposing the system of requirements into subsystems, and these themselves into further subsystems, in such a way that each subsystem contains a set of requirements very densely connected internally, yet as far as possible independent of the requirements in other subsystems.

To do this, we think of each requirement as a binary variable, capable of being in two states (satisfied and unsatisfied). The links between the requirements are interpreted as probabilistic tendencies for the linked variables, to take equal or opposite values. With such an interpretation, where the village's environment appears as a system of interdependent binary variables, the task of finding the environment's subsystems becomes the task of decomposing the set of variables into subsets of variables in such a way that some function expressing the total dependence between the sets is minimized.

To make this clear, I first give the simplest possible way of doing it. If we divide the set of linked points into two sets, some of the links will probably be severed. Obviously if such a partition severs a great many links, then the variables in one set are very much intertwined with those in the other, and the partition is not a good one. On the other hand, if the partition severs fewer links, the two sets are less dependent. If we treat the number of links severed as the criterion function, and minimize this function, we shall have two sets of variables which we might reasonably call independent subsystems.

As it happens, this particular function is over-simple and cannot be justified theoretically. I shall now describe two functions which can be derived from reasonable theoretical premises.

The first function is derived from the problem of establishing components as well-integrated units. It is a measure of the amount of information transmitted by one set of variables to the other:

$$\frac{(\ell - \ell_a - \ell_b)m(m-1)/2. - \ell.a.b}{\sqrt{\{a.b.(m(m-1)/2 - a.b)\}}} *$$

If we partition the variables in such a way as to minimize this function, we get subsystems which are informationally as independent as possible. The derivation and exact purpose of this function are given in my thesis. Minimization according to this function has been programmed for the IBM 7090. It is this function which gave the decomposition of the village problem that follows.

The second function is derived from the problem of having new needs affect no more than one component at a time. Suppose that in the future some new need arises, not at present represented in the set of variables. If the present city is to adapt to this new need successfully, we should hope that only one of the existing city's components need be modified in response to the new need. In other words, we should hope that the existing subsystems are such that this new requirement be linked entirely to those requirements in some one subsystem, and not at all linked to any of the requirements in other subsystems.

Assume that any new requirement is linked to a random selection of the existing requirements, subject to the following condition: A link between two existing requirements makes it slightly more likely that a new requirement will be linked to both of them or neither, and slightly less likely that it will be linked to just one of them. We can then express the total probability of a new requirement being linked to just one subsystem, not to both, for any division of the existing requirements into two subsystems. This probability is given by:

$$2^b\left(\ell_a - \frac{a(a-1)}{m(m-1)}.\ell\right) + 2^a\left(\ell_b - \frac{b(b-1)}{m(m-1)}.\ell\right)$$

$$\sqrt{\left\{2^{2b}.\frac{a(a-1)}{2}\left[\frac{m(m-1)}{2} - \frac{a(a-1)}{2}\right] + \right.}$$

$$+ 2^{2a}\frac{b(b-1)}{2}\left[\frac{m(m-1)}{2} - \frac{b(b-1)}{2}\right]$$

$$\left. - 2^{a+b}\frac{a(a-1)}{2}.\frac{b(b-1)}{2}\right\}$$

---

*The parameters used in the functions are to be interpreted as follows:
m the total number of variables
a the number of variables in one subsystem
b the number of variables in the other subsystem
ℓ the total number of links
ℓ_a the number of links entirely within the first subsystem
ℓ_b the number of links entirely within the other subsystem

We partition the variables in such a way as to maximize this probability. I am now in the course of programming this maximization for the IBM 7090.

IV.   EXAMPLE

Analysis of the graph defined by the 141 village needs and the links between them, shows that these needs fall into four major subsystems which I have called A, B, C, D; and that these systems themselves break into twelve minor subsystems, A1, A2, A3, B1, B2, B3, B4, C1, C2, D1, D2, D3. Each of these minor subsystems contains about a dozen needs.

A1 contains requirements 7, 53, 57, 59, 60, 72, 125, 126, 128.

A2 contains requirements 31, 34, 36, 52, 54, 80, 94, 106, 136.

A3 contains requirements 37, 38, 50, 55, 77, 91, 103.

B1 contains requirements 39, 40, 41, 44, 51, 118, 127, 131, 138.

B2 contains requirements 30, 35, 46, 47, 61, 97, 98.

B3 contains requirements 18, 19, 22, 28, 33, 42, 43, 49, 69, 74, 107, 110.

B4 contains requirements 32, 45, 48, 70, 71, 73, 75, 104, 105, 108, 109.

C1 contains requirements 8, 10, 11, 14, 15, 58, 63, 64, 65, 66, 93, 95, 96, 99, 100, 112, 121, 130, 132, 133, 134, 139, 141.

C2 contains requirements 5, 6, 20, 21, 24, 84, 89, 102, 111, 115, 116, 117, 120, 129, 135, 137, 140.

D1 contains requirements 26, 29, 56, 67, 76, 85, 87, 90, 92, 122, 123, 124.

D2 contains requirements 1, 9, 12, 13, 25, 27, 62, 68, 81, 86, 113, 114.

D3 contains requirements 2, 3, 4, 16, 17, 23, 78, 79, 82, 83, 88, 101, 119.

We may picture this system of subsystems as a tree (Figure 1).

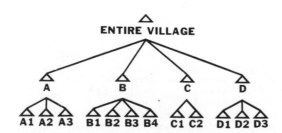

ENTIRE VILLAGE

A B C D

A1 A2 A3 B1 B2 B3 B4 C1 C2 D1 D2 D3

Figure 1

I repeat what I have said before. If we regard this system of needs as the environment, and we regard the village as a physical system of components constantly striving for adaptation to its environment, then we want to put the hierarchy of the village's components, in 1–1 correspondence with the hierarchy of subsystems in the environment. It is under these circumstances that expansion and modification of the village are likely to be most successful. To create this 1–1 correspondence, I have tried to establish, for each of the twelve subsystems of needs, a physical component which meets first the needs in that subsystem, and no other needs. These components are shown in diagrammatic form in the following pages. I have also tried, in the four major diagrams labelled A, B, C, D (Figures 2–5), to show how the minor components might fit together to make the major components, and then in the composite diagram (Figure 6) how A, B, C, D themselves might fit together to make an entire village.

I first give a summary of the components, and the way they fit together, so that the more detailed account of each component and the functions which belong to it may be better understood.

The four main components are roughly these: A deals with cattle, bullock carts, and fuel; B deals with agricultural production, irrigation, and distribution; C deals with the communal life of the village, both social and industrial; D deals with the private life of the villagers, their shelter, and small-scale activities. Of the four, B is the largest, being of the order of a mile across, while A, C, D are all more compact, and fit together in an area of the order of 200 yd across.

The basic organization of B (Figure 2) is given by the component B4, a water collector unit, consisting of a high bund, built in the highest corner of the village, at right angles to the slope of the terrain; within the curve of this bund, water gullies run together in a tank. This tank serves the rest of the village land, which lies lower, by means of sluices in the bund; the component B4 is intimately connected with B3, the distribution system for the fields. The principal element of this component is a road elevated from floods, which naturally takes its place along the top of the bund defined by B4. At intervals along this road, distribution centres are placed providing storage for fertilizer, implements, and seeds; in view of the connection with B4, each one of these centres may be associated with a sluice, and with a well dug below the bund, so that it may also serve as a distribution centre for irrigation water. Each distribution centre serves one unit of type B2; this is a unit of cooperative farming broken into contoured terraces by anti-erosion bunds and minor irrigation channels running along these bunds. B1 is a demonstration farm surrounding the group of components ACD, just at those points of access which the farmers pass daily on their way to B2 and B3.

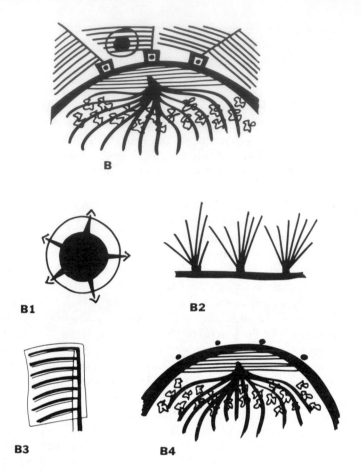

Figure 2

The smaller group of components ACD is given its primary
organization by the fact that several units of type D must
function together (Figure 3). Each D copes with the small-scale
activities of about 50 people. It is defined by D2, a compound
wall carrying drinking water and gas along its top. At the
entrance to the compound, where the walls come together, is a
roofed area under which cottage industries take place. The
compound contains the component D1, an assembly of storage
huts, connected by roofed verandahs which provide living
space. Every third or fourth hut has a water tank on top, fed by
the compound wall, and itself feeding simple bathing and
washing-up spaces behind walls. D3 is a component attached to
the entrance of the compound; it provides a line of open water
at which women may wash clothes, trees with a sitting platform
at their base for evening gossip, in such a way that the water and
trees together form a climatic unit influencing the microclimate
of the compound, and also, because of the water and trees,
offering a suitable location for the household shrine.

C (Figure 4) is made of two components; C2 is a series of
communal buildings (school, temple, panchayat office, village

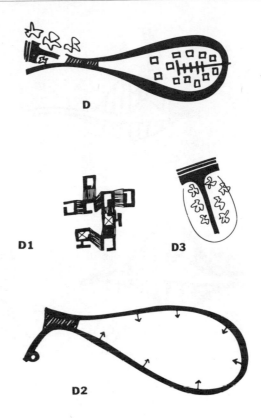

Figure 3

meeting place, etc.) each with a court, the courts opening in alternatingly opposite directions. The cross walls are all pierced by gates, in such a way that there is a continuous path down the middle of the component. This path serves as a connecting link between different centres, a processional route, and pedestrian access to the compounds D which may therefore be hung from C2 like a cluster of grapes. One end of this component C2 runs into C1; C1 is a widening of the road on the bund; on this widening out, a number of parallel walls are built to mark out narrow, urban-like plots. There is in the centre of these plots a bus stop, opening out of the road itself. The whole unit houses whatever industry, power sources, and other aspects of the village's future combine base, develop.

The structure of A (Figure 5) starts with A2, a group of cattle stalls, each stall opening towards the outside only, its floor falling towards the centre, with a drain in the centre leading all manure to a pit where the slurry for the gober gas plant can be prepared. Each compound has such a component A2 in its centre, between the pieces of D1; exit from the compound, for cattle and carts, is by way of component A3, a gate in the compound wall, marked by the cattle trough and the gober gas plant itself. A group of several components A2 and A3 are tied together by the single A1. A1 consists of a central control point through which all cattle leaving any compound, have to pass.

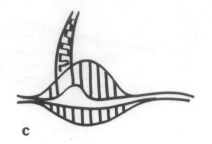

C

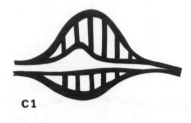

C1

C2

Figure 4

A

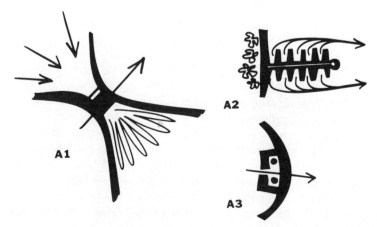

A1

A2

A3

Figure 5

This control point provides a hoof bath, a dairy, and a link to the main road via C1.

There now follows a more detailed account of the reasons behind the organization of each of the twelve minor components.

A1:   7. Cattle treated as sacred and vegetarian attitude.
     53. Upgrading of cattle.
     57. Protection of cattle from disease.
     59. Efficient use and marketing of dairy products.
     60. Minimize the use of animal traction to take pressure off shortage.
     72. Prevent famine if monsoon fails.
    125. Prevent malnutrition.
    126. Close contact with village level worker.
    128. Price assurance for crops.

The sacredness of cattle (7) tends to make people unwilling to control them, so they wander everywhere eating and destroying crops, unless they are carefully controlled. Similarly, the need to upgrade cattle (53) calls for a control which keeps cows out of contact with roaming scrub bulls; and further calls for some sort of centre where a pedigree bull might be kept (even if only for visits); and a centre where scrub bulls can be castrated. Cattle diseases (57) are mainly transferred from foot to foot, through the dirt—this can be prevented if the cattle regularly pass through a hoof bath of disinfecting permanganate. If milk (59) is to be sold cooperatively, provision must be made for central milking (besides processing); if cows are milked at home, and the milk then pooled, individual farmers will adulterate the milk. Famine prevention (72), the prevention of malnutrition (125), and price assurance for crops (128) all suggest some kind of centre offering both storage, and production of nourishing foods (milk, eggs, groundnuts). If

Figure 6   **ENTIRE VILLAGE**

the village level worker (126) is to come often to the village and help, quarters must be provided for him here. Animal traction (60) calls for access to and from the cattle stalls (A2) on the one hand, and the road on the other.

A2: 31. Efficient distribution of fertilizer, manure, seed from village storage to fields.
     34. Full collection of natural manure (animal and human).
     36. Protection of crops from thieves, cattle, goats, monkeys, etc.
     52. Improve quantity of fodder available.
     54. Provision for feeding cattle.
     80. Security for cattle.
     94. Provision for animal traffic.
    106. Young trees need protection from goats, etc.
    136. Accommodation of wandering caste groups, incoming labour, etc.

Here (31, 34, 54, 80, 94) form a subset connected with cattle movement and manure, while (36, 52, 106, 136) form a subset mainly concerned with the protection of crops and trees from wandering cattle. (31) and (34) call for the collection of urine and dung, which suggests cattle should be in one place as much of the time as possible, where there is a pucca floor draining towards a central manure collector. This is of course closely connected with feeding stalls, the most permanent standing place for cattle. (80) calls for psychological security—cattle owners want their cattle as near to them as possible, if not actually in the house, and therefore absolutely opposed to the idea of a central communal cattle shed. In view of disease and germ breeding difficulties the closest arrangement possible seems to be one where individual stalls are immediately opposite owners' verandahs with nothing but a path between; this path serves to accommodate cattle traffic (94). Each stall is marked by its walls, roofed only by wood purlins at 2 ft centres, so that the fodder itself, stored on top, provides shade. Rains are not heavy enough to warrant permanent roofing. Vegetables, young trees, etc., which would be specially benefited by protection from cattle, must either be very far away, or else very close so that separation can really be achieved by a barrier (36, 196). To make this work, (52) must be assured by other means—stall feeding perhaps, which then connects with (54). To prevent the cattle of wandering shepherds causing trouble (136), the proper grazing ground must abut the road, the access to it must be the normal road–village access. This grazing ground should be on the good land side of the bund, so that when green silage is introduced, land can be irrigated and cultivated.

A3: 37. Provision of storage for distribution and marketing crops.
    38. Provision of threshing floor and its protection from marauders.
    50. Protected storage of fodder.
    55. Cattle access to water.
    77. Village and individual houses must be protected from fire.
    91. Provision and storage of fuel.
    103. Bullock cart access to house for bulk of grain, fodder.

Access for cattle to water (55) should be to good water, hence to drinking water distribution system, feeding off compound wall D2. (77) and (91) are best achieved by a controlled fuel supply, like gas, supplied by a gober gas plant using manure from A2, the gas distributed to individual kitchens by the same artery that distributes water, i.e. the compound wall.

At the point on the compound wall indicated by these previous items, there must be an opening to allow passage of bullock carts (103), and at this point there should also be a store for supplies and fodder—or at least an easy unloading and access point to the roofs of the cattle bays (37, 38, 50).

B1: 39. Best cotton and cash crop.
    40. Best food grain crop.
    41. Good vegetable crop.
    44. Crops must be brought home from fields.
    51. Improve quality of fodder available.
    118. Demonstration projects which spread by example.
    127. Contact with block development officer.
    131. Panchayat must have more power and respect.
    138. Achieve economic independence so as not to strain national transportation and resources.

(39, 40, 41, 51) and economic independence (138) are all items which can only be improved by the widespread use of improved agricultural methods; these are not directly dependent on the physical plan, but on a change of attitude in the villagers. This change of attitude cannot be brought about by sporadic visits from the agricultural extension officer and village level worker, but only by the continuing presence of demonstration methods, on site, (118); there should be a demonstration farm, government or panchayat owned (131), perhaps run by the village level worker in association with the panchayat (hence accommodation for such officers, 127). (118) and (44) suggest that the farm be placed in such a way that every farmer passes it daily, on his way to and from the fields.

B2: 30. Efficient and rapid distribution of seeds, fertilizer, etc., from block HQ.

35. Protection of crops and insects, weeds, disease.
46. Respect for traditional agricultural practices.
47. Need for new implements when old ones are damaged, etc.
61. Sufficient fluid employment for labourers temporarily (seasonally) out of work.
97. Minimize transportation costs for bulk produce (grain, potatoes, etc.).
98. Daily produce requires cheap and constant (monsoon) access to market.

(97) and (98) are critical, and call for access to and from the fields on a road which is not closed in the monsoon; i.e. on an embankment. (30) and (35) call for efficient distribution within the plots, of seeds, fertilizers, insecticides, etc., which must themselves be stored at some point where delivery is easy—i.e. on the road. Hence the idea of distribution centres located at intervals along the main road, serving wedge-shaped or quasi-circular units of agricultural land. (46, 47, 61) have little discernible physical implication.

B3: 18. Need to divide land among sons of successive generations.
19. People want to own land personally.
22. Abolition of Zamindari and uneven land distribution.
28. Proper boundaries of ownership and maintenance responsibility.
33. Fertile land to be used to best advantage.
42. Efficient ploughing, weeding, harvesting, levelling.
43. Consolidation of land.
49. Cooperative farming.
69. Fullest possible irrigation benefit derived from available water.
74. Maintenance of irrigation facilities.
107. Soil conservation.
110. Prevent land erosion.

(18–49) all point to the development of cooperative farms of some sort, from the point of view of increasing efficiency of resources, manpower, machines, better crops, rotation of crops, etc. (69) cannot be implemented unless water is distributed from the HQ of such cooperatives because otherwise faction and personal rivalries, etc., prevent full use of wells— i.e. warring neighbours adjacent to the source of water (well) will not agree to cooperate about sharing its use. (74) irrigation requires consolidated ownership of channels, otherwise neglect at one place holds up the efficient use somewhere else. (107) soil conservation depends on rotation of crops, which is only feasible if large plots are under single ownership control, so that they can carry the full pattern of rotation. (110) erosion is prevented by long continuous contour bunds, which can only be put across land of integrated ownership. Bund and irrigation divisions on contours suggest terraced strips of land as units of cooperative farm, fed from single uphill source.

B4: 32. Reclamation and use of uncultivated land.
45. Development of horticulture.
48. Scarcity of land.
70. Full collection of underground water for irrigation.
71. Full collection of monsoon water for use.
73. Conservation of water resources for future.
75. Drainage of land to prevent waterlogging, etc.
104. Plant ecology to be kept healthy.
105. Insufficient forest land.
108. Road and dwelling erosion.
109. Reclamation of eroded land, gullies, etc.

(32) and (48) call for use of wasteland, which often contains river bed area. (48) calls for irrigation of this area. (71, 73, 75) suggest the use of monsoon water instead of and as well as well water for irrigation, since well irrigation is temporary in the long run, because it causes a drop in the water table. Apart from actually using monsoon water for irrigation, the water table in the wells can be preserved if the wells are backed up by a tank. Hence a curved bund, collecting water above wells placed under the bund (70). Rainfall in the catchment area (again a water resource issue (73)) will be improved by tree planting (104, 105) which suggests putting fruit trees (45) inside the curve of the bund. (Incidentally, placing the trees within the bund offers us a way of protecting young trees from cattle, by keeping the cattle on the other side of the bund, which then forms a natural barrier.) Further, if water is to flow toward tank, horizontal contour bunds cannot be used to check erosion as they are in B3, so erosion of gullies, streams, etc., can only be controlled by tree planting (109). Road erosion is controlled if the road is on top of the bund itself (108).

C1:   8. Members of castes maintain their caste profession as far as possible.
10. Need for elaborate weddings.
11. Marriage is to person from another village.
14. Economic integration of village on payment in kind basis.
15. Modern move towards payment in cash.
58. Development of other animal industry.
63. Development of village industry.
64. Simplify the mobility of labour, to and from villages, and to and from fields and industries and houses.
65. Diversification of village's economic base—not all occupations agricultural.

66. Efficient provision and use of power.
93. Lighting.
95. Access to bus as near as possible.
96. Access to railway station.
99. Industry requires strong transportation support.
100. Bicycle age in every village by 1965.
112. Access to a secondary school.
121. Facilities for birth, pre- and post-natal care (birth control).
130. Need for increased incentives and aspirations.
132. Need to develop projects which benefit from government subsidies.
133. Social integration with neighbouring villages.
134. Wish to keep up with achievements of neighbouring villages.
139. Proper connection with bridges, roads, hospitals, schools proposed at the district level.
141. Prevent migration of young people and harijans to cities.

This is composed of two major functional sets: (11, 64, 95, 100, 112, 121, 133, 134, 139) which concerns the integration of the village with neighbouring villages and with the region, and (8, 10, 14, 15, 58, 63, 65, 66, 93, 96, 99, 130, 132, 141) which concerns the future economic base of the village, and all the aspects of 'modern' life and society.

These two are almost inseparable. They call for a centre, away from the heart of the village, on the road, able because of being on the road, to sustain connections between the village and other villages (11) and capable of acting as a meeting place for villagers of different villages (112, 121). This function is promoted by the need to provide a bus stop (95), village industries with optimum access to the road (63–66, 99), the social gathering place connected with the bus and with jobs made available by the industries (133, 134, 61); the development of a modern and almost urban atmosphere to combat migration of the best people to cities (141), and to develop incentives (14, 15, 130, 132). A centre of industry to promote (8, 63, 64). The road satisfies (64, 95, 96, 99, 100, 139). The centre will be the natural physical location for sources of power and electricity transformer (66, 93); also the most efficient place for the poultry and dairy farming which require road access (58); the bus stop is the natural arrival place for incoming wedding processions (10).

C2:   5. Provision for festivals and religious meetings.
       6. Wish for temples.
      20. People of different factions prefer to have no contact.
      21. Eradication of untouchability.

24. Place for village events—dancing, plays, singing, etc., wrestling.
84. Accommodation for panchayat records, meetings, etc.
89. Provision of goods, for sale.
102. Accommodation for processions.
111. Provision for primary education.
115. Opportunity for youth activities.
116. Improvement of adult literacy.
117. Spread of information about birth control, disease, etc.
120. Curative measures for disease available to villagers.
129. Factions refuse to cooperate or agree.
135. Spread of official information about taxes, elections, etc.
137. Radio communication.
140. Develop rural community spirit: destroy selfishness, isolationism.

The major fact about the communal social life of the village is presence of factions, political parties, etc.; these can be a great hindrance to development (20, 129). If the various communal facilities of the village (5, 6, 24, 84, 89, 111, 115, 120, 137) are provided in a central place, this place will very likely get associated with one party, or certain families, and may actually not contribute to social life at all. On the other hand, it is important from the point of view of social integration (21, 140) to provide a single structure rather than isolated buildings. What is more, isolated buildings also have possible connection with the single family nearest them, which can again discourage other families from going there. What is required is a community centre which somehow manages to pull all the communal functions together, so that none is left isolated, but at the same time does not have a location more in favour of some families than others. To achieve this, a linear centre, containing some buildings facing in, some out, zigzagging between the different compounds is necessary. This also meets (102) the need for processions with important stopping places; and adult literacy calls for a series of walls along the major pedestrian paths, with the alphabet and messages written in such a way that their continuing presence forces people to absorb it (116, 117, 135).

D1: 26. Sentimental system: wish not to destroy old way of life. Love of present habits governing bathing, food, etc.
29. Provision for daily bath, segregated by sex, caste, and age.
56. Sheltered accommodation for cattle (sleeping, milking, feeding).
67. Drinking water to be good, sweet.

76. Flood control to protect houses, roads, etc.
85. Everyone's accommodation for sitting and sleeping should be protected from rain.
87. Safe storage of goods.
90. Better provision for preparing meals.
92. House has to be cleaned, washed, drained.
122. Disposal of human excreta.
123. Prevent breeding germs and disease starters.
124. Prevent spread of human disease by carriers, infection, contagion.

Houses, as they are used at present, are chiefly store-rooms; people actually live on their verandahs most of the time. The one thing which inner rooms provide, namely privacy and psychological security, appears among the needs to be met by D2, not here. Hence, we solve (87) by providing store-rooms, which in a column like manner support verandah roofs stretching between them (85). (26) is mainly concerned with bathing and food, connected with (67, 29, 90) these suggest a water store on top of occasional store houses, with kitchen and bath wall attached to this store (also 122); probably this water store will be fairly close to the source of water as we shall see when we combine this with D2. (76) the floor of the verandah must be raised to keep it out of flood water—also the compound should drain towards the centre to remove the dangers of (92, 123, 124). (56) calls for a space to house A2.

D2: 1. Harijans regarded as ritually impure, untouchable, etc.
    9. Members of one caste, like to be together and separate from others, and will not eat or drink together with them.
   12. Extended family is in one house.
   13. Family solidarity and neighbourliness even after separation.
   25. Assistance for physically handicapped, aged, widows.
   27. Family is authoritarian.
   62. Provision of cottage industry and artisan workshops and training.
   68. Easy access to drinking water.
   81. Security for women and children.
   86. No overcrowding.
  113. Good attendance in school.
  114. Development of women's independent activities.

(1, 9, 12, 13) suggest group compounds, as they are found at present, each of about 5 to 10 families, i.e. 25 to 50 persons. To provide security (81), especially for women surround it by a wall, whose top serves as a distribution channel for water (68). The fact that the space within the wall is all protected, allows women more freedom within the compound for women's communal activities (114), gives more freedom to widows (25), and allows cottage industries, which are likely to be run largely by women, to flourish (62). The space for cottage industry (62) should go at the entrance to the compound, where women going to and fro from washing activities pass it constantly; this may to some extent combat the effects of purdah (27); encourges women to come out from their houses (which the rooms of a usual house discourages, because it allows women to shut themselves up in seclusion), and may even help girls' attendance in school by making the women more bold (113). Since containing walls are moved outward, overcrowding is less likely to take place (86)—adjustment and expansion can take place more easily within the compound walls, than within individual house walls.

D3:   2. Proper disposal of dead.
        3. Rules about house door not facing south.
        4. Certain water and certain trees are thought of as sacred.
     16. Women's gossip extensively while bathing, fetching water, on way to field latrines, etc.
     17. Village has fixed men's social groups.
     23. Men's group chatting, smoking, even late at night.
     78. Shade for sitting and walking.
     79. Provision of cool breeze.
     82. Provision for children to play (under supervision).
     83. In summer people sleep in open.
     88. Place to wash and dry clothes.
    101. Pedestrian traffic within village.
    119. Efficient use of school; no distraction of students.

Here there are several overlapping functions. (23, 78, 79, 82, 83) all require the control of climate—in particular getting cool conditions—which can be best achieved by the juxtaposition of water and trees. (16, 17, 23, 88, 101) require a unit for gossip, washing clothes, meeting purposes, at the compound level. (2, 3, 4) demand the construction of a place with certain qualities of sacredness, perhaps quiet, water, neem trees. Pedestrian traffic and quiet are called for again by (101, 119). All these functions call for a unit in which water, trees, washing facilities, pedestrian movement, sitting under the trees are juxtaposed; the unit fits directly onto the compound, just outside the entrance. Washing may be either on ghats, etc., or on steps fed from the water wall unit D2.

In conclusion I should like to emphasize the following points:
1. The diagram I have shown for the entire village is only one way of putting the components together. There are many many ways of doing it. The way they fit together in any

actual case will depend on local peculiarities of site and population.

2. It is not at all necessary that these components be introduced into an existing village structure all at once. In fact, although the components do function well in company with one another, each one of the twelve is capable of being introduced into the fabric of an existing village by itself. It is just this that I mean by calling these components the proper units of development.

REFERENCES

The theory this paper is based on is to be found in Christopher Alexander, "Notes on the synthesis of form", Ph.D. thesis, Harvard University, 1962.

The computer program for decomposing a graph, according to the function described on page 41, is written up in Christopher Alexander and Marvin Manheim, *Hidecs 2; A Computer Program for the hierarchical decomposition of a set with an associated linear graph*, Civil Engineering Systems Laboratory Publication No. 160, MIT June 1962.

1.3 Systematic Method for Designers

L. Bruce Archer

INTRODUCTION

The traditional art of design—that is, selecting the right material and shaping it to meet the needs of function and aesthetics within the limitations of the available means of production—has become immeasurably more complicated in recent years. While user needs were simple, materials few, and manufacturing methods relatively crude, the designer was able to adopt rules of thumb to meet them. Within the boundaries of these limitations, the art of industrial design was something close to sculpture.

Today, the designer is faced with subtler evidence of user needs and market demands. He is presented with a galaxy of materials to choose from, many of them having no true shape, colour, or texture of their own. The means of production have become more versatile, so that rules of thumb provide no easy guide. At the same time, the cost of tooling frequently means that the designer (or his manufacturer) cannot afford to be wrong. Paradoxically, the relaxation of the limitations of materials and processes has made the job of the designer more difficult, rather than easier, since he must now choose and decide in many cases where the decision was previously made for him.

In the face of this situation there has been a world-wide shift in emphasis from the sculptural to the technological. Ways have had to be found to incorporate knowledge of ergonomics, cybernetics, marketing, and management science into design thinking. As with most technology, there has been a trend towards the adoption of a systems approach as distinct from an artefact approach.

Originally published by The Design Council, London (1965). Reproduced by permission of The Design Council.

The most fundamental challenge to conventional ideas on design, however, has been the growing advocacy of systematic methods of problem solving, borrowed from computer techniques and management theory, for the assessment of design problems and the development of design solutions. The following text describes some of these developments and sets out one method for organizing the design act systematically. [...]

1. THE NATURE OF DESIGNING

Before we can look at systematic methods for designers, we must know what we mean by 'design'. An architect preparing plans for a house is clearly designing. So is a typographer preparing a layout for a page of print. But a sculptor shaping a figure is not. What is the difference? A key element in the act of designing is the formulation of a prescription or model for a finished work in advance of its embodiment. When a sculptor produces a cartoon for his proposed work, only then can he be said to be designing it. Sometimes the word 'creating' is employed when there is no model or prescription between the formation of the idea and its embodiment. Thus a couturier may 'create' a gown directly into its final form, but if the gown is a model gown its 'creation' is the 'designing' of the line.

So, the prior formulation of a prescription or model is one necessary element. The hope or expectation of ultimate embodiment as an artefact is another. Hence the formulation of the idea for an office filing system may be designing, so long as it anticipates the laying down of 'hardware'. Similarly, the discovery of a chemical formula in general is not designing, but the prescription of a formula for (say) a new plastics material may be. Thus, we say that not only a prescription or model, but also the embodiment of the design as an artefact, are essential to the definition of designing. In this sense the composition of music, for example, although in many ways analogous, is not designing.

There is also a sense in which the act of arriving at a solution by strict calculation is not regarded as designing. Here, the solution can be seen to arise automatically and inevitably from the interaction of the data, and the process of calculation is regarded as non-creative. It is characteristic of creative solutions (and often of the most successful designs) that they are seen to be apt solutions—but *after* completion and not before. Some sense of originality is also essential. No-one would accept that the preparation of new drawings for a perfectly standard clothes peg could be graced with the title of designing. However, just how much originality is needed to distinguish the preparation of working details from actual designing, and just how little inevitably is required to distinguish calculation from designing, is very difficult to proscribe, especially in the field of mechanical, electronic, and structural engineering. The satisfaction of these two conditions together is sometimes referred to as the presence of a creative step.

So we find that design involves a prescription or model, the intention of embodiment as hardware, and the presence of a creative step. This definition embraces the central activities of architecture, most forms of engineering (including some systems engineering), certain sciences, all industrial design, and most applied art and craft. It implies a purposeful seeking after solutions rather than idle exploration. It also implies that certain limitations exist and that recourse to any random action is not enough.

There can be no solution without a problem; and no problem without constraints; and no constraints without a pressure or need. Thus, design begins with a need. Either the need is automatically met, and there is no problem, or the need is not met because of certain obstacles or gaps. The finding of means to overcome these obstacles or gaps constitutes the problem. If solving the problem involves the formulation of a prescription or model for subsequent embodiment as a material object (and requires a creative step), then it is a design problem. The skills required for its solution depend upon the nature of the predominating constraints. These determine whether the problem is called architecture, engineering, applied science, industrial design, or art and craft. All these terms are more or less vague in their comprehensiveness, and tend to overlap or merge into one another at their fringes. Anything that is said about design applies as much to one as to the other, but as this paper is being presented in an industrial design context, the remainder of the argument will be conducted from an industrial design point of view.

The term 'industrial design' is generally used to cover a class of design problems which ranges from domestic appliances, office machinery, and public service equipment on the one hand to typography, textiles, and wallpaper on the other. All industrial design is characterized by the importance of the visual element in the design of the end product and by the fact that the end product is made or reproduced by industrial methods. The visual element may be purely aesthetic, as in the design of a public interior, or frankly commercial, as in advertising display or product styling. The industrial methods of production may be repetitive, as for domestic appliances or print; or one-off, as for exhibitions and displays. (Hand-made pottery and jewellery, if subject to prescriptions or models in advance of embodiment, are included in our general definition of design but, because they are made by the hand of the designer, for our purposes they fall into the sub-class of craft rather than into the general field of industrial design.)

The art of reconciliation

We have already said that the art of designing is the art of
reconciliation. There is a whole complex of factors arising from
the three main aspects of industrial design—function, marketing, and manufacture. These factors are always competing and
sometimes they are in direct conflict, but they must be
reconciled in the end product. The word 'compromise' is
avoided, because it implies some half-way house which might
very well leave each of the competing requirements equally
unsatisfied. Reconciliation implies that the conflict is resolved.

At a later stage we shall try to identify systems of interacting
factors as they occur in particular problems. For the moment, it
might be useful to consider the classes into which these factors
generally fall. Take the point of use, for example. Design begins
with a need. The product is a means for fulfilling that need. It is
a tool. It performs some work (Figure 1). Psychologists tell us
that most human work is aimed at stabilizing the environ-

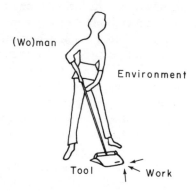

Figure 1
Design begins with a need. The
product is a means for fulfilling
that need. It is a tool.

ment—we build houses to keep in a consistent climate, and to
keep out predators. We grow, gather, store, cook, and eat food
to keep our internal metabolism on an even keel. When the
summer sun shines and our bellies are full, we like to lie on the
grass and sleep. It is when a cold wind blows or our blood sugar
falls that we are impelled to get up and so some work.

The current tendency in design, as in many other fields, is to
try to consider the whole system of which the proposed
product is a part, instead of considering the product as a
self-contained object. The basic system at the point of use may
be described as a man–tool–work–environment system (Figure
2). The activities implied by this system are shown in Figure 3
and listed in Table 1. For convenience, the considerations
involved can be reduced to three human factors (motivation,
ergonomics, aesthetics) and three technical factors (function,
mechanism, structure).

Examination of similar generalized conditions at the point of
manufacture and the point of presentation or sale suggests three

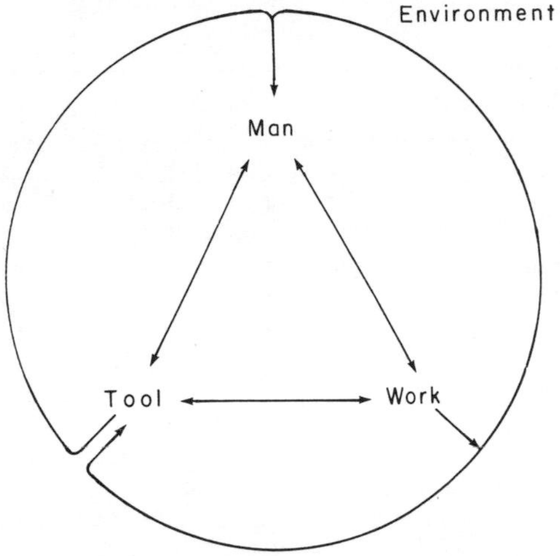

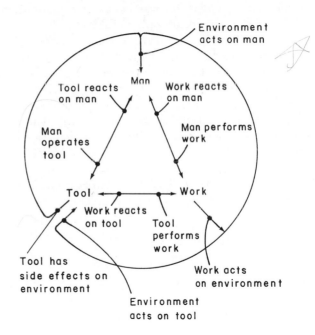

Figure 2
The basic relationship at the point
of use.

Figure 3
The activities contained in a
man–tool–work environment
system.

Table 1
Activity

1	Environment acts on man
2	Man performs work
3	Work reacts on man
4	Man operates tool
5	Tool reacts on man
6	Tool performs work
7	Work reacts on tool
8	Work acts on environment
9	Environment acts on tool
10	Tool has side effects on environment
11	Environment acts on man

Table 2

Class of design factors	Relevance		
Aesthetics	Use	Sale	
Motivation	Use	Sale	
Function	Use	Sale	
Ergonomics	Use		
Mechanism	Use		Manufacture
Structure	Use		Manufacture
Production		Sale	Manufacture
Economics		Sale	Manufacture
Presentation		Sale	

further factors (production, economics, presentation), indicating nine design factors altogether (Table 2).

Of course, these lists are neither exhaustive nor exact, but the tables form a useful basis for the evolution of checklists and analytical systems. The most important thing to observe about the factors listed is the difference between their fundamental qualities. Some, such as structure and economics, relate to matters of fact susceptible to measurement which can be optimized by conventional methods of calculation. Others, such as motivation and aesthetics, relate to matters of value which can only be assessed on the basis of a kind of case law. This variation in the quality of the factors is characteristic of architectural and industrial design problems.

A complex of sub-problems

In practice, of course, the designer cannot define the factors in his particular problem purely in terms of these lists. A single design problem is a complex of a thousand or more sub-problems. Each sub-problem can be dealt with in a characteristic way—by operational research, working drawings, value-judgments, etc. But although each sub-problem can be resolved so as to produce an optimum solution, or even a field of acceptable solutions, the hard part of the task is to reconcile the solutions of the sub-problems with one another. Often, where the optimum solution of one sub-problem compels the acceptance of a poor solution in the other, the designer is forced to decide which of the two must take priority. This entails putting the whole complex of sub-systems into an order of importance, often referred to as 'rank ordering'. Rank ordering is an activity at which human beings can become extremely skilled, given sufficient experience. In fact, it is this particular skill that practitioners of the various professions are largely paid for. Rank ordering of dissimilar factors is very difficult to work out by logical or mathematical methods, however. It is only when there is little experience to go on (as in sending rockets to the planets), or when the consequences of being wrong are disastrous (as in building an atomic reactor) that it becomes worthwhile even to try. Thus, although resolving a large number of sub-problems and listing all their combinations and permutations is the very thing that computers are good at, it is unlikely that any computer will replace the designer in the role of criterion-giver or judgment-maker—at least for a very long time to come.

Why, then, are we bothering to talk about systematic methods of designing? The fact is that being systematic is not necessarily synonymous with being automated.

Computer science is a very good means for setting out a problem in a systematic and logical way. In the design field it can truthfully be said that, having prepared a problem for a computer, the answer quite often will become so obvious to a skilled designer that he can dispense with the computer's services altogether. It will not always be certain that the answer he gets is the best or the only answer, but since he is very rarely asked to *prove* that his is the best answer, if it works it will usually suffice.

Main point of criticism

The most devastating opposition to the application of systematic methods to design problems comes from those who point out that the results of analysis are usually ponderous statements of the blindingly obvious. Perhaps it is worth remembering that it is also characteristic of *creative* designs that they are seen to be apt solutions—*after* they are made. A solution which in retrospect seems obvious is not necessarily to be condemned. Systematic methods come into their own under one or more of three conditions: when the consequences of being wrong are grave; when the probability of being wrong is high (e.g. due to lack of prior experience); and/or when the number of interacting variables is so great that the break-even point of man-hour cost versus machine-hour cost is passed. The last-mentioned situation has already been reached in the design of industrial plant and large hospitals. For lesser problems the routines worked out for rigorous analysis act as checklists of those procedures which the designer must carry out, however intuitively, and the checklists help to ensure that everything is taken into account.

Unfortunately, the science of design method has not yet reached a degree of sophistication which will permit the use of agreed axioms, or even the use of an agreed terminology. The several scattered research workers in this field each have their own favourite models, techniques, and jargon. However, a certain amount of common ground is emerging. For example, a basic breakdown of the nature of design procedure is largely

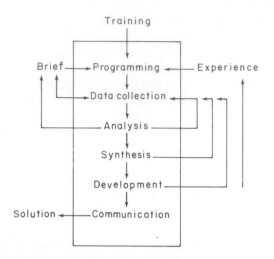

Figure 4
A breakdown of basic design
procedure.

agreed, although there are some differences about whether it
should be described in three stages, or four, or six. The present
author favours six. The stages and their interrelation are shown
in Figure 4. In practice, the stages are overlapping and often
confused, with frequent returns to early stages when difficulties are encountered and obscurities found.

One of the special features of the process of designing is that
the analytical phase with which it begins requires objective
observation and inductive reasoning, while the creative phase at
the heart of it requires involvement, subjective judgment, and
deductive reasoning (Figure 5). Once the crucial decisions are
made, the design process continues with the execution of
working drawings, schedules, etc. again in an objective and
descriptive mood. The design process is thus a creative
sandwich. The bread of objective and systematic analysis may
be thick or thin, but the creative act is always there in the
middle.

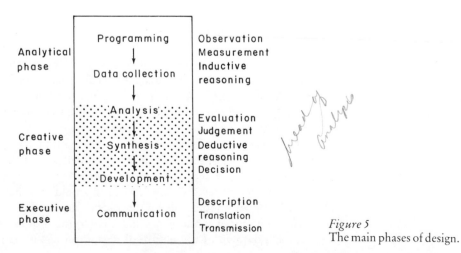

Figure 5
The main phases of design.

Summary

It has been seen, then, that design is defined as involving a prescription or model, the intention of embodiment as hardware, and the presence of a creative step. It has been said that industrial design is especially characterized by the importance of appearance in the end product, and by the fact that this end product is reproduced by industrial methods. It has been argued that the art of (industrial) designing is essentially the art of reconciling a wide range of factors drawn from function, manufacture, and marketing. It has also been argued that the process of designing involves analytical and creative, objective and subjective phases. The practice of design is thus a very complicated business, involving contrasting skills and a wide field of disciplines. It has always required an odd kind of hybrid to carry it out successfully. The more sophisticated the demands of function and marketing become, and the more versatile the available materials and processes become, the harder the job of the designer will get. Already it has become too complicated for the designer to be able to hold all the factors in his mind at once.

2. GETTING THE BRIEF

An old recipe which begins 'First catch your hare ...' is often quoted as an example of beginning at the beginning. Exemplary as the sentiment may be, however, the model is not perfect. Catching the hare is not the beginning. There must first be awareness of the need. This awareness is based upon the recognition of a cyclical pattern of hunger and action, frustration and satisfaction. A history of the previous occasions when hunger was experienced, and of the consequences of various prudent and imprudent actions which have been taken in the past, is stored, recalled, and used as a basis of future action. This is the way in which, in nature, day-to-day problems are solved. Biologists tell us that highly organized response mechanisms operate something like this:

1 discern 'something wrong'
2 heighten alertness and locate the area of the disturbance
3 bring appropriate sense organs to bear
4 evaluate evidence and compare with previous experiences
5 postulate a possible cause for the disturbance
6 recall experience with similar and/or analogous causes
7 predict consequence of suggested cause
8 formulate courses of action possible in response to such a cause
9 recall experience with similar and/or analogous courses of action
10 predict consequences of each suggested course of action

11   select course of action to be followed
12   act
13   discern effect of action
14   recall experience with similar and/or analogous effects
15   repeat from 7 until equilibrium is restored.

The study of the control mechanisms of living organisms is
called cybernetics. In recent times, designers of highly complicated control systems for machine tools, aeroplanes, rockets,
and remote-controlled instruments have turned to cybernetics
for inspiration. Such has been their success that biologists now
look at computers and control-and-feedback mechanisms for
explanations of how living organisms might work. Computer
programming often follows the 'biological' sequence of problem solving very closely. Paradoxically enough, because computers are really such quick but stupid beasts, rather advanced
mathematics has had to be employed to permit problems to be
expressed in simple enough terms. These simple terms have
been seized by research workers in several fields. They, like the
biologists, are now using the ideas as inspiration in non-
computer applications. It is in this roundabout fashion that
some of the techniques to be described in this paper have come
into play.
    A further line of thinking which does not quite fall into this
pattern, but which has contributed to the development of
systematic methods for designers, is the 'heuristic', an ancient
philosophical study of the method of intellectual discovery
which has been revitalized recently by Professor G. Polya of
Stanford University, USA (Polya, 1945). Although Professor
Polya is a mathematician, his heuristic is concerned with
plausible rather than exact reasoning. We use plausible rather
than exact reasoning in everyday life, because the evidence
available to us is itself scrappy and inexact. When we hear the
cry, 'stop thief!' and see a man running, we guess that he is the
culprit, but the reasoning which leads us to this conclusion is far
from foolproof. Yet it gives us enough of a cue to take actions
which may permit us to confirm or deny our assumption later
on. Professor Polya emphasizes that, although plausible
reasoning can and does produce solutions to problems, it
cannot be represented as proof. If proof is needed it must be
worked out retrospectively. The method for solving design
problems set out in this paper owes something to both the
heuristic and the cybernetic approaches.

Asking the right questions

There are plenty of good designers who have no difficulty at all
in producing the right answers, if only they are asked the right
questions. Most designers, good and bad, find that the
problems they are asked to solve are seldom clearly defined by

their clients. A client may be in stage 1 of our cybernetic sequence (discern something wrong), and may really need a management consultant rather than a designer. He may be in stage 3 (bring appropriate sense organs to bear), and the designer may be the sense organ thought appropriate. In any case, so far as the client's problem is concerned, the designer starts to operate at stage 4 (evaluate evidence and compare with previous experience). But for the moment, this examination is directed to the opening task of a design problem, getting the brief.

In Section 1, 'design' was defined as a goal-seeking activity, so the next step after focusing on the task is to identify the goals. The client will have altruistic as well as commerical goals: there may be social aims imposed by the community; the designer may have professional and personal aims and ambitions of his own. When these conflict, another set of ethical problems arises. For the moment all the aims and ambitions must be looked on as equal and relevant. Design is necessarily concerned with change. For our purposes, identifying the goals means defining the needs and pressures which constitute the driving force for change. This force may be generated by the pull towards certain rewards, or it may be the product of repulsion from certain penalties. In any event its resultant points out the direction called 'good', so far as progress in the solution of that particular problem is concerned (Figure 6).

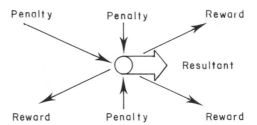

Figure 6
The driving force for change may be the result of many pressures and needs.

Identifying constraints

The next task is to identify the constraints which hedge about this force for change. Constraints there must be, or there would be no problem. Sometimes the constraints will be spread out, providing the boundaries of a field of manoeuvre (Figure 7). This is the 'open' design situation, where many solutions are feasible. The direction pointed out by the goal-seeking forces indicates the corner of the field of freedom in which design solutions should be sought. In other cases the field of manoeuvre may be so small that the required solution is virtually conditioned by the interaction of the constraints (Figure 8). This is the inevitable or mathematical solution. Too often the constraints overlap, one required condition com-

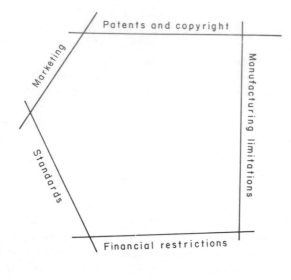

Figure 7
The design constraints mark out
the boundaries of a field for
manoeuvre.

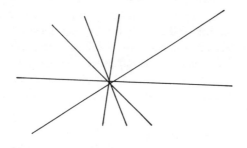

Figure 8
The field for manoeuvre may be so
small as to allow only a single
solution.

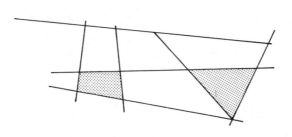

Figure 9
Often the superimposed
constraints provide a broken-up
field of freedom, suggesting two or
more kinds of design solution.

pletely denying the fulfilment of another. In this case the brief
must be re-appraised and the client asked to relax or remove one
of his conditions altogether. If this is not possible, then a
technical (or other) breakthrough may be required to produce
means for reconciling the overlapping constraints.

Very often the superimposed constraints provide a broken up
field of freedom (Figure 9), in which two or more fundamentally different design prescriptions are indicated. The identifica-

tion of goals and constraints corresponds to Professor Polya's opening questions:

'What is the unknown?'
'What is the condition?'
'Is it possible to satisfy the condition?'
'Is the condition sufficient to determine the unknown?'
'Or redundant?'
'Or contradictory?'

Crucial issues

Rarely, if ever, is this information handed to the designer in predigested form. Preliminary analysis of the given data is necessary to complete the brief. The cybernetic model suggests that the designer should now postulate the cause (or root) of the problem, compare with previous experience, and suggest, as a first approximation, a course of action which might be taken.

Seekers after a rigorously logical design method may protest that the solution can only come out of analysis of *all* the data, and that the postulation of a possible course of action at this stage is premature. This argument presupposes the possibility of obtaining full and exact information on every relevant fact. The argument is also, in itself, the postulation of a particular course of action (analysis of all the data before hypothesis). In real-life problems, perfect information is rarely available. Making a first approximation on the basis of prior experience enormously reduces the scale of the problem solving effort.

Accordingly, the checklist method proceeds to identify the crucial issues and a course of action for resolving them. Professor Polya says, 'Find the connection between the data and the unknown. Have you seen it before? Do you know a related problem?' The collection of case histories and a review of previous experience play vital roles in formulating possible courses of action in the approach to a design problem. An experienced man, working in a field he knows, draws upon his skill and judgment without conscious analysis. Where previous experience is limited (as when the designer is new to the field, or when the field itself is newly developed), a more formal analysis may be necessary. Where the penalty for error is grave (as in life-or-death equipment) a rigorous examination is required.

Seeking answers

So far, the technique described has involved little or no data analysis, in the normal sense of the words. This is because the purpose of the opening phase is to define the problem and formulate a course of action. Most designers must submit an outline programme with estimates of costs for approval before the work of evaluating data and developing solutions can be undertaken.

To have defined the problem properly—even to have put a finger on the crucial issues—is not the same as having solved the problem itself. Nevertheless, it has gone some way toward a solution and, having formulated some sort of plan, the designer can offer estimates of time and cost. Assuming that his programme is approved, he can then proceed to find answers to the questions he has posed.

3.   EXAMINING THE EVIDENCE

It is axiomatic that any rational method for solving design problems must offer means for arriving at decisions on the basis of evidence. A wholly logical method would demand clearly defined goals, rigorous decisions, full feedback, and perfect information. 'Clearly defined goals' mean that both the direction called 'good' and the boundaries of acceptability are known in respect of every situation requiring decision. 'Rigorous decisions' means following the course of action which leads to optimum results in terms of the given goals. 'Full feedback' means that, after each decision, all previous decisions are reappraised in the light of the later decision. 'Perfect information' means that all the facts are known, all the facts are accurate, and all the factors taken into account are relevant.

In the case of an interplanetary instrument package extremely accurate prediction of performance is needed, and must be achieved within close limits. A rigorous solution is both desirable and feasible. In the case of most industrial design problems a rigorous solution is not feasible because of the lack of perfect information, and is undesirable because of the high cost of carrying out full feedback. Moreover, both the user's and the producer's situation demand or permit quite a loose fit between the product's characteristics, its performance requirements, and the limits of its materials. Any practicable method of designing must take this into account. The techniques described in this paper, though based on a hypothetical, rigorous method, are therefore calculated to allow plausible reasoning and inexact information.

Finding the information

The paucity of reliable information about the factors which must be taken into account in the design of a product is the bugbear of the industrial designer. Several methods have been suggested to help designers to collect economically what information there is. J. Christopher Jones of Manchester University (Jones, 1963) advocates beginning with a joint session of all the people concerned with the design project. Each person lists those matters which his experience or imagination tell him are relevant to the problem in hand. These 'random factors', as Mr Jones calls them, are recorded on cards. No attempt is made to avoid overlapping or to eliminate wild ideas. Indeed, no comment or criticism on the part of the other participants is allowed. Each suggestion, however, triggers off new thoughts in everyone else's mind, and the process is continued until an apparently exhaustive list of the random factors has been made. Samples, photographs, drawings, etc., may be introduced in order to assist the exploration.

This technique is a variant on the familiar brainstorming idea, and it is well calculated to extract in the shortest possible time a great deal of the knowledge embodied in the experience of all the people concerned. According to Mr Jones' method, the cards are then sorted into categories and the lists of factors are used as checklists to guide the collection of data from appropriate sources. (An interaction chart is useful in checking where one factor is affected by others, so as to identify sub-problems.)

Some practitioners prefer to work from a basic diagram, or from a hierarchy or thesaurus of factors. Others prepare checklists of the factors they have found by experience to be relevant to particular types of problem.

Dealing with the information

Having gathered whatever information is available, the next question is how to deal with it. Most information is empirical, reliably true only for a given situation at a given time. Market research reports, laboratory reports, specifications of existing products, photographs, etc., all fall into this class. For the purposes of design, this information must be converted to a form that can be used to define the required characteristics of the proposed new product, or to define the limitations of the means available to produce, market, and use it. The creative act of designing will be to find an apt manner of providing those characteristics within those means. The immediate need, however, is to organize the information so that the designer can make useful decisions.

Evaluation and decision

Accordingly, the next step is to prepare, on the basis of the checklist and information collected, a list of all the matters which require evaluation and decision. Matters which are interdependent are paired or grouped so as to form a series of sub-problems. It is necessary at this stage to distinguish between those sub-problems which relate to the characteristics required of the proposed new product (i.e. problems about ends, described as P-specs by Mr Jones) and those sub-problems which relate to materials and processes (i.e. problems

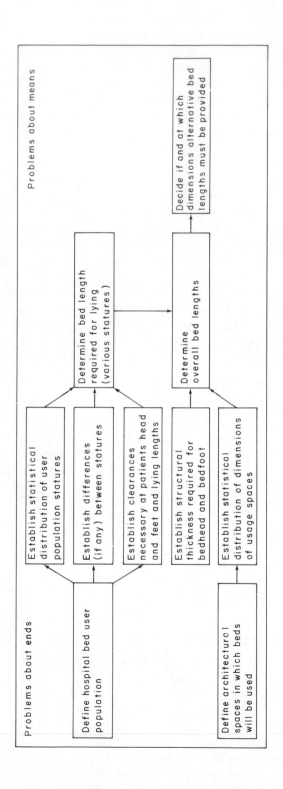

Figure 10
Example of an arrow diagram. The
information necessary for the
solution of one problem often
hangs on the outcome of another.

about means). The way in which the sub-problems depend upon one another may be demonstrated by means of an arrow diagram, (Figure 10). The arrow diagram helps to ensure that decisions are made in the right order, and also provides a ready guide to the feedback of one decision to another.

Different problems demand different techniques. Some problems can be solved by simple arithmetic, and some by technical drawing. Others may need advanced mathematics or bench tests or opinion sampling. There is a technique appropriate to every problem.

The next step is to put all the sub-problems into a rank order of significance. The interdependence of sub-problems has already been noted. The new rank order is in terms of precedence, if it should prove that achievement of the optimum in one sub-problem is incompatible with the optimum of another (Figure 11). For example, if the optimum finish for a teapot interior (in terms of resistance to staining and corrosion) should prove to be incompatible with the optimum selling price of the teapot (in terms of ruling market prices and manufacturing costs), then which must take precedence, the best finish or

Figure 11
Table showing the theoretical range of relationships between the solutions to two interacting problems. The field of mutually compatible solutions is represented by shaded areas. The optimum solution in each case is represented by a + in the centre of the circle. The best situation is shown in 5, where the optima coincide. When this is not the case, as in 4, 3 and 2, a solution *x* may be chosen within the shaded area but nearest to the optimum which is considered more important, A or B. In the case 1, no mutually compatible solution is possible and the constraints must be reconsidered.

THERE ARE FIVE DEGREES OF COMPATIBILITY OF THE SOLUTION OF ANY TWO RELATED SUBPROBLEMS	
Diagrammatically	In words:
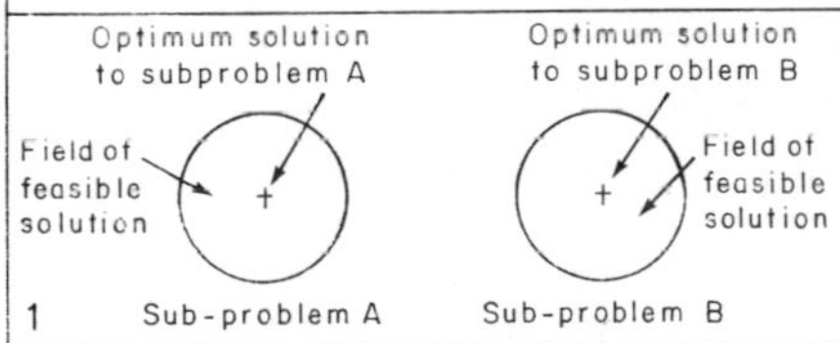 Optimum solution to subproblem A — Optimum solution to subproblem B — Field of feasible solution — Field of feasible solution — 1 — Sub-problem A — Sub-problem B	Wholly incompatible solutions: there is no overall solution which embraces feasible solutions to *both* sub-problems
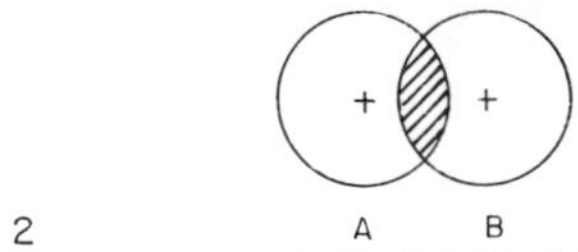 2 A B	Mutually incompatible optima: there is no overall solution which embraces an optimum solution to either sub-problem, but there *are* overall solutions which embrace mutually feasible solutions to both sub-problems (see note above)
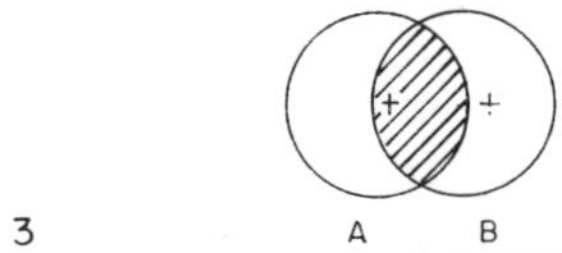 3 A B	Unilaterally incompatible optimum. there is no overall solution which embraces the optimum solution to sub-problem B, but there *are* overall solutions, including one which provides the optimum solution to sub-problem A (see note above)
 4 A B	Alternate optima: there are overall solutions, which may include the optimum solution to A or B but not to both simultaneously
 5 A B	Coincidental optima: there are overall solutions including one which satisfies the optima of both sub-problems simultaneously

Rank ordering chart: a hypothetical case						
A	B→					

Rank ordering chart: a hypothetical case

A	B→ The factors referred to by these numbers correspond with those in the left hand column					
↓	1	2	3	4	5	score
1 Optimum materials	·	0	0	0	0	0
2 Optimum use of plant	1	·	1	1	1	4
3 Optimum finish	1	0	·	0	0	1
4 Optimum selling price	1	0	1	·	0	2
5 Optimum sales volume	1	0	1	1	·	3

Figure 12
Example of a ranking chart.
Priorities are decided pair by pair,
and the overall ranking is
mathematically determined.

the best selling price? The use of a binary chart, which is a simple and effective device for rank ordering, especially where more than one judge is involved, is demonstrated (Figure 12). The consequence of this, and the succeeding reappraisal, is the assembly of all the facts about the characteristics desired of the new product, and all the facts about the limitations of the means for producing, marketing, and using it, into a performance specification.

A brief recap

In essence, the technique described in this paper so far has five main stages:

(1) The aims and objects of the design effort are determined, together with the essential criteria by which a 'good' solution will be distinguished from a 'not so good' solution.
(2) The factors affecting the design are identified and listed.
(3) The ways in which factors depend upon, or interact with, one another are established. Pairs of groups of dependent or interacting variables are designated 'sub-problems'.
(4) The factors or sub-problems are arranged in an order of priority to indicate which of any pair of sub-problems should take precedence if it should prove impracticable to provide the 'best' solution to one sub-problem (say, the best component material for durability purposes) at the same time as the 'best' solution to another sub-problem (say, the best component material for ease of manufacture).
(5) Each sub-problem is dealt with by appropriate means.

Finally,

(6) The whole design problem is expressed as a rank ordered list of the attributes which the final solution is required to have. This much might well be computerized.

The result, however, is a statement of the problem, not of the answer.

4.   THE CREATIVE LEAP

When all has been said and done about defining design problems and analysing design data, there still remains the real crux of the act of designing—the creative leap from pondering the question to finding a solution. In our original examination of the nature of designing (Section 1), we stipulated the presence of a creative step as an essential element, distinguishing design from certain other problem solving activities. We also defined industrial design as necessarily containing aesthetic and other human value-judgments. If we accept that value-judgments must be made by people, and that value-judgments cannot be the same for all people, for all places, or for all times, then it follows that neither the designer nor his client (nor, eventually, the user) can abdicate the responsibility for setting up his own standards. Similarly, there is no escape for the designer from the task of getting his own creative ideas. After all, if the solution to a problem arises automatically and inevitably from the interaction of the data, then the problem is not, by definition, a design problem.

Codifying the central task

Thus any systematic method for designers must allow for the establishment of human values at certain points, and for the introduction of original ideas at others. Although in some ways it is a contradiction in terms to try to codify creativity, it might be desirable to do so at this point if only to see whether or not we are right in saying that the designer's central task cannot be mechanized.

Much has been written on the mechanism of the creative leap, mainly in the field of scientific discovery. For centuries, the ideal of science had been the suppression of guesswork and the substitution of logic. However, the advent of the quantum theory and wave mechanics in nuclear physics marked the end of the concept of science as the uncovering by deductive logic of the immutable laws of nature. Science having discovered nuclear effects which were causeless and random, statistical methods of gauging probability took over from the search for hidden laws. Kant had forecast this 100 years earlier when he wrote 'The law of gravitation (i.e. that attraction varies inversely with the square of the distance) is arbitrary—God could have chosen another.'

Today, we can suspect that all the laws of nature are arbitrary, or even that there are no laws at all. For example, it can be argued that *if* the universe has existed for all time, extending indefinitely in all directions through limitless space, and *if* random aggregations and mutations are occurring everywhere, then it follows that sooner or later, somewhere or other, a planet like Earth was bound to be present and a form of life such as evolutionary animal life culminating in man was bound to occur.

For those who prefer the 'big bang' theory, this analogy hardly holds, but nevertheless, out of boundless time we can suspect that phenomena which on our tiny time scale look remarkably like fixed 'laws of nature' may well be mere coincidences—moments in randomly changing relationships.

Max Planck remarked: 'We have no right to assume that any physical laws exist (with respect to some phenomenon under examination) or, if they existed up to now, that they will continue to exist in a similar manner into the future' (Planck, 1931).

The creative mechanism

With scientists taking this sort of view of science, designers should be unembarrassed at accepting the transience of design. Human values, fashion, and public taste may well be describable in the same terms, if not on the same time scale, as probabilistic phenomena in physics. So may the mechanism of the creative leap. Thus we can assert that if enough people think hard enough about a problem for long enough, somebody, somewhere, will hit on an apt and original solution. This is (or was) the way of the arts. An abundance of artists starving in garrets is the surest guarantee of an artistic breakthrough.

The technique of brainstorming in design is based upon the same principle. Here a group of people, well informed on various aspects of the problem in hand, allow their fantasies to run riot, triggering off in one another's minds a torrent of ideas which no amount of logic could have produced. Or could it? It has been argued that all mental processes are goal-directed and predictable (Mace, 1964), and capable of being imitated by a computer (whereas ERNIE, the Post Office computer which picks the premium bonds winning numbers, specializes in random behaviour). Computers are better and quicker than human beings at the chore of trying every single permutation of the facts. Therefore one can assume that, in certain circumstances, a computer might be able to do better than a brainstorming team, or than a designer—provided, of course, that someone stands by to spot the bright idea when it comes out or, alternatively, instructs the machine on the criteria for identifying apt and original solutions for itself. This brings us back to our original point—setting up goals and criteria and spotting bright ideas are tasks which cannot be abdicated.

Searching for solutions

Even so, it is worth considering how the creative act is carried out. A designer searches in his mind for a solution to his problem by examining all kinds of analogies. He looks mainly at other people's end results, checking whether something on those lines would answer his own problem. Only when he has reviewed all sorts of solutions, including phenomena and artefacts in the most unlikely fields, does he return to the question and examine other questions of a comparable kind handled by himself and others. If this still yields no result, he tries to reformulate the problem. Only as a last resort does he attempt deductive reasoning, proceeding from the analysis of the data to the necessary conclusion, instead of the other way round.

But since it is the judgment of scientific philosophers that outside mathematics the ideal of deductive reasoning is largely illusory, there can be nothing unscientific about the traditional reliance on intuition and inspiration in design.

There is one important lesson to be learned before we leave the subject of the creative leap, however. The transactional school of perception theorists has demonstrated with startling clarity the effect of prior experience, expectation, and purpose on man's capacity to perceive evidence and judge hypotheses. For sheer survival, man has developed an agile machinery for filtering out, from the hail of signals with which his senses are bombarded during every waking moment, those few which are significant at any given time.

The filtering apparatus of the nervous system suppresses what it takes to be accidental, spurious, or irrelevant sensations (which may be of internal as well as external origin) before these reach the consciousness. Where necessary, the subconscious may supply what it takes to be the most likely materials to fill any blanks still remaining. A. Ames Jnr., W. H. Ittelson, and F. P. Kilpatrick at Princeton University, and M. W. Perrine at Ulm, have shown that what we *think* we see is thus based upon comparison with complex collections of previous experiences and previous expectations, fulfilled or disappointed.

The startling and crucial conclusion is that we *cannot* believe our own eyes, and that the most painstaking care has to be taken, both in data analysis and in hypothesis seeking, to counterbalance the effects of the perceptual filter. The term 'transactional theory of perception' implies that the observer contributes from his own experience, by either addition or subtraction, to his perception of the phenomenon before him. Perception is a two-way business. We are thus brought face to face with the reality of the need for rich, wide, and fruitful experience among designers, as well as the capacity for flexibility and fantasy in thought.

One major contribution that systematic methods of designing might make, especially when supported by mechanical aids, is to reduce the dull, imagination-suppressing chores which the designer now has to undertake, releasing him to devote more of his time to equipping himself for his crucial task—that of making the creative leap. And this he must do alone.

5.   THE DONKEY WORK

The creative phase of design has been most systematically developed in Britain in the design departments of art schools. The development phase is the great strength of the technical college. Yet it is architects who practise the most conscious and systematic progression from the one phase to the other, through schemes, sketch plans, ⅛ in. plans, ½ in. details, full-size details, and renderings. It is architects, too, who most consciously recognize the difference between work which is calculated to test the feasibility of a design idea and work which is intended to convey to the production department instructions for an embodiment of the design. Whether or not architects are better than engineers at actually doing it is another question, and we shall come to that later.

Section 4 referred to an analysis of the designer's mental processes in the act of synthetics—the conceiving of the basic idea for a solution to his problem. It is characteristic of mental processes that they work in loops, taking one jump ahead then two steps back with great rapidity. The anticipatory development work in the synthesis phase to which we have referred was intended to test the appropriateness of one or more design ideas as answers to the functional problem in the brief, and to test the feasibility of one or more design ideas from the practical point of view. The detailed development in the development phase, however, is intended to fill gaps and solve problems in making a selected design idea work. Indeed, for commercial purposes it is usually necessary to suppress any new thoughts on basic design ideas once the development phase has been entered. When severe snags are encountered at a late stage in development, it is often better to seek changes in the design brief or in the production constraints than in the design idea itself.

The idea and its embodiment

There is a real distinction between a design idea and any one embodiment of it. The design idea, if it fits our definition of the answer to a design problem (Section 1) also fits both the dictionary and the patent law definitions of an invention. On the other hand, the finished design, being an embodiment of the design idea rather than the design idea itself, may be subject to copyright rather than to letters patent. Even in a patent application, the inventor is asked to describe separately the invention and a material embodiment of it. The description of the invention is interpreted literally, and is deemed to cover all the variants that the inventor wishes to cover. The description of the material embodiment of the invention is interpreted freely and is regarded merely as an exemplar. Hence, there is a logical difference both in intention and in method between the kind of development work and feasibility testing which goes on in the synthesis phase, and that which goes on in the development phase. Whereas abstract analysis is needed to prove that a given design idea is the best recipe for an occasion, the proof of the design pudding is still in the eating.

In some industries—furniture, for example—it is often quicker and cheaper to build a prototype and submit it to user tests that it is to carry out extensive detailing and stress calculation. The advantage of suck-it-and-see methods is that, however subtle the variables, a direct measure of overall success or failure is possible. The chief disadvantage is that so many problems of construction must be wholly or partly solved before performance testing can even begin.

In most industries, however, the preliminary development work is carried out on paper. Mechanical and structural engineers, for example, often carry the paperwork to the final acceptance of the smallest detail before a single component is made. Sometimes this gets out of hand. A team of hundreds of draughtsmen and stressmen, exhaustively defining and cross-checking thousands of design elements, may begin to regard the determination of dimensions and tolerances on detail drawings as the 'real' work, and the ultimate assembly and test of a prototype as 'the answer to last week's puzzle'.

Methods of design development

Engineers may be weak in the systematic construction of the brief and in searching for original design ideas, but they are strong in the techniques of the development of detail. As a development tool, the orthographic arrangement drawing would take a lot of beating. But the scale drawing is no more than a model or analogue which closely represents the geometry of the proposed design, while being quicker to prepare and modify than the real thing.

Drawings are not the only analogues which can be used in design development. Three-dimensional models can be made in different materials or to a reduced or enlarged scale in order to check appearance or convenience, or for wind-tunnel purposes. Stress models can be made for photo-elastic and strain-gauge analysis. Electrical and/or mathematical models of facets other than size and shape can be erected for breadboard or computer analysis. The more abstract the model, the more basic and flexible can the appraisal be. The more realistic the model, the more directly can overall or ultimate effects be judged.

The lesson, then, is to use the more abstract of the means available for development and test of basic design ideas, and to use the more realistic of the means available for development and test of the design embodiment. Table 3 shows the range of means by which a design idea may be expressed.

There is another lesson to be learned, too. When the designer has finished with his designing, other people usually take over. Production engineers, buyers, and tool designers have additional matters to take into account. To give them sufficient room to manoeuvre, the product designer should strive to keep to a minimum the inter-dependence between design elements, and to a maximum the ability of any one design element to accept coarseness or variance. One production engineer has said that the best design is the crudest that will just do the job. A production-orientated designer might say that the best design is that which gives the highest-quality whole with the lowest-quality elements.

Sometimes the detailed development work is carried out by, or under the direction of, the designer. Sometimes it is done by the production side. In either case there is no escape from the

Table 3

Type of analogue	Examples	
1 A form of words	Evocative words Definitive statements Patent specifications	When testing *basic design idea*, use most abstract model which will fully express it→
2 Symbolic logic	Boolean algebra Mathematical models	
3 Diagrams	Flow diagrams Circuit diagrams Vector diagrams	
4 Sketches	Evocative sketches Definitive sketches	
5 Formal drawings	Perspectives Renderings Scale drawings	←When testing *a design embodiment* use most concrete model which can be afforded
6 Simple models	Block models Space models Scale models	
7 Working analogues	Electrical analogues Rigs Photoelastic models	
8 Prototype		

donkey work of defining the position; orientation; shape; size; structural qualities; colour; weight; and chemical, electrical, or other physical properites of every single element in the design.

However, the information available to the designer is so patchy or unreliable on so many matters that it would be impossible to arrive at solutions without making some assumptions or judgments which go beyond the evidence. Almost any proposed design solution constitutes a hypothesis based upon imperfect evidence, and it must be subjected either to the test of the market place or to some indirect analysis. The development phase cannot be considered complete until validation studies have been carried out.

REFERENCES

Jones, J. C. (1963), 'A method of systematic design', in Jones, J. C. and Thornley, D. (eds), *Conference on Design Methods*, Pergamon, Oxford.

Mace, C. A. (1964), *The Psychology of Study*, Penguin, Harmondsworth.

Planck, M. (trans. W. H. Johnston) (1931), *The Universe in the Light of Modern Physics*, George Allen and Unwin, London.

Polya, G. (1945), *How to Solve It*, Princeton University Press, Princeton, New Jersey.

APPENDIX: Checklist for product designers

Summary

0	Preliminaries	
0.1	Receive enquiry	
0.2	Evaluate enquiry	
0.3	Estimate office workload	
0.4	Prepare preliminary response	
1	Briefing	Phase 1—'Receive brief,
1.0	Receive instructions	analyse problem, prepare
1.1	Define goals	detailed programme and
1.2	Define constraints	estimate'
2	Programming	
2.1	Establish crucial issues	
2.2	Propose a course of action	
3	Data collection	Phase 2—'Collect data, pre-
3.1	Collect readily available information	pare performance (or design) specification, reap-
3.2	Classify and store data	praise proposed programme
4	Analysis	and estimate'
4.1	Identify sub-problems	
4.2	Analyse sub-problems about ends	
4.3	Prepare performance specification	
4.4	Reappraise programme and estimate	

5	Synthesis	Phase 3—'Prepare outline
5.1	Resolve problems about ends	design proposal(s)'
5.2	Postulate means for reconciling divergent desiderata in specification	
5.3	Develop solutions-in-principle to problems about means arising from specification	
5.4	Postulate outline overall solution(s)	
6	Development	Phase 4—'Develop proto-
6.1	Define design idea	type design(s)'
6.2	Erect a key model	
6.3	Develop sub-problem mutual solutions	
6.4	Develop overall solution(s)	
6.5	Validate hypotheses	Phase 5—'Prepare (and ex-
7	Communication	ecute) validation studies'
7.1	Define communication needs	Phase 6—'Prepare manufac-
7.2	Select communication medium	turing documentation'
7.3	Prepare communication	
7.3	Prepare communication	
7.4	Transmit information	
8	Winding up	
8.1	Wind up project	
8.2	Close records	

[In the original there follows a fuller, considerably expanded checklist and activity flowcharts.]

1.4 An Approach to the Management of Design

John Luckman

INTRODUCTION

Design is a man's first step towards the mastering of his environment. The act of designing is carried out in a wide variety of contexts including architecture, town planning, industry, engineering, arts and crafts.

So far operational research has contributed very little to existing design methodology and this paper is offered as a possible starting point.

Our introduction to the subject was by way of a study of the process of design as seen by architects. We have subsequently extended our early ideas to cover two of our present research projects involving engineering design and city planning.

The definition of design we have used, following Fielden *et al.* (1963) is: 'The use of scientific principles, technical information and imagination in the definition of a structure, machine or system to perform prespecified functions with the maximum economy and efficiency.' This definition is, if anything, a little more bold than others in that it includes scientific principles from the beginning.

Designing has become more difficult than it was. In a rapidly changing technological world, the number of materials, the variety in size, shape and colour of parts or the whole, the variation in manufacturing methods, the range of qualities that are required, the refinement of tolerances that are acceptable and the pressure on performance of the finished article have all

Originally published in *Operational Research Quarterly*, **18** (4) (1967), 345–58. Reproduced by permission of Pergamon Press Ltd.

combined to make the designer's job more complex. This has led towards an almost universal search for systematic methods to assist in unravelling the problems that these factors pose.

However, as our research progresses, it becomes apparent that a study of the design process on its own is not sufficient, since the majority of pressures on the designer are external to it. To understand the limitations, constraints and objectives of the design process it is necessary to know more of the research and development process of which design is a part. Within this larger process, design needs to be managed. This paper puts forward some simple models of the design process which we hope will eventually be enlarged to cover the management of design within a wide range of research and development processes.

THE PROCESS OF DESIGN

The process of design is the translation of information in the form of requirements, constraints, and experience into potential solutions which are considered by the designer to meet required performance characteristics. Here we shall insist that some creativity or originality must enter into the process for it to be called design. If the alternative solutions can be written down by strict calculation, then the process that has taken place is not design.

Research workers in the field of design have not yet agreed upon a single set of axioms or even terminology in their models of the design process. Some ground that is common, however, is that the design process can be described in terms of a number of stages.

In this paper we will be using three stages—other authors (Asimow, 1962; Jones, 1966) use three, four, or six. The stages are:

(1) *Analysis*: The collection and classification of all relevant information relating to the design problems on hand.
(2) *Synthesis*: The formulation of potential solutions to parts of the problem which are feasible when judged against the information contained in the analysis stage.
(3) *Evaluation*: The attempt to judge by use of some criterion or criteria which of the feasible solutions is the one most satisfactorily answering the problem.

In practice the whole process of design will consist of many levels, progressing from very general considerations at the start, through to specific details as the project nears completion. In architecture, for example, the process will start with the general problems of spatial layout, site access, amenity, etc., and progress through problems of room relationship and treatment of exterior elevations until at the level of final working drawings, the problem of floor finishes, decorations, and minor items such as door handles, lighting fitments, etc., arise.

Thus, a level is synonymous with a sub-problem within the total problem where a set of interconnected decisions must be taken, and at every level to a greater or lesser degree, the stages of analysis, synthesis, and evaluation are used.

For the majority of the earlier levels, the stage of evaluation is performing the function of an indicator of satisfactory areas for development at the next level. Any decisions made become part of the input to the next level and naturally situations can arise when it is better not to single out one feasible solution, but rather keep several ideas open to allow more thorough exploration of the next level. By doing so, the designer is leaving open the chance of using feedback to the earlier levels.

The broad overall model does not conflict with common sense since the majority of design problems cannot be totally comprehended without taking the logical step of trying to break them up into sub-problems which are as independent as possible. The total solution is then built up from the solutions to the sub-problems, step by step.

However, Alexander (1964) argues strongly for a careful reconsideration of the way in which sub-problems are defined. He contends that sub-problems are commonly defined using concepts that are on occasion arbitrary and unsuitable, and generally without full regard to the underlying structure of the total problem. Because of this, the sub-problems themselves are highly interconnected and offer no savings or short cuts in reaching the total design solution.

We feel this point is highly relevant and we discuss this further later.

ONE LEVEL OF THE DECISION PROCESS

What is the designer or design team doing when taking decisions within any one particular level of the design process? The flow diagram (Figure 1) represents our interpretation of the technical content of one level of the design process.

The groups, G, represent sets of information collected at the outset of the project. As indicated this comes from many sources, including the information from previous levels as the result of earlier decisions and information which is the individual designer's experience. Sometimes the information is not formally written down (and indeed sometimes cannot be written down). Our interpretation of the analysis stage is one of collecting, classifying, and collating this information to a greater or lesser degree depending upon the designer. In this sifting-and-sorting activity the designer is seeking the creative step that will take him on to the next stage, synthesis.

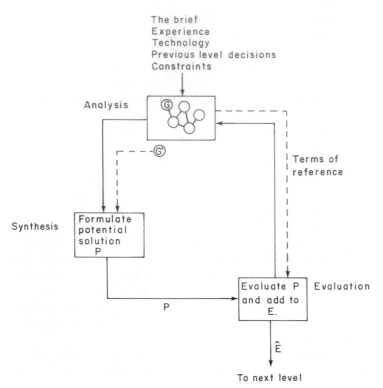

The brief
Experience
Technology
Previous level decisions
Constraints

Analysis

Synthesis

Formulate
potential
solution
P

Terms of
reference

Evaluate P   Evaluation
and add to
E.

P

Ê

To next level

Figure 1
One level of the design process.

   The design process develops by formulating a potential
solution, *P*, which is a result of the creative step just taken,
checking its constituent parts against the information contained
in *G*, and if required, information freshly collected as in *G'*.

   The designer will proceed only if his potential solutions is
compatible with that information. Each potential solution is
taken to the stage of evaluation where it will be tried and tested
against the criteria that the designer or the process dictates. If
accepted, it will add to or replace some or all of the previously
evaluated potential solutions, *E*. The cycle of analysis, synthesis, and evaluation is repeated until the level is completed by
the acceptance of one or more of the evaluated potential
solutions (*Ê*) say, and this signals a move to a new level of the
total problem.

   Now it is clear that this concept of what is happening within a
level of the design process takes longer to write down in many
cases than it will take to carry out in practice. The designer does
not consciously take every step laboriously but, like a good
chess player, can see several moves ahead and back, taking
short cuts, and is often checking feasibility of an idea at one
level with ideas for other levels. Nevertheless, we feel that the
technical content of what the designer is trying to do is
represented fairly accurately by the model.

SYSTEMATIC METHODS

Can help be given to a designer or a design team faced with such a technical problem? Research work so far suggests several systematic methods that can assist designers.

It should first be made clear that being systematic does not mean that the designer is superfluous with solutions being generated by a computer. The principal area where systematic methods have been tried is in the analysis stage. Orderly and logical presentation of basic data is a great advantage in helping the designer to make his creative step to a potential solution, especially in complex design problems. Systematic analysis comes into its own when the number of information groups is large, when the consequences of coming up with a poor or wrong solution are expensive or wasteful and when the chance coming up with a poor or wrong solution is high, i.e. when there is lack of prior experience in the field.

Some of the simplest systematic techniques that are commonly used are in the form of 'checklists'. These checklists set out the information groups which the designer ought to have thought about, and while, of course, they do not ensure that the designer has the right thoughts, at least the particular item has not been completely forgotten. A typical method is that of J. C. Jones of Manchester University who advocates the recording of 'random factors' by each member of the design team who are encouraged to write down anything that occurs to them to be relevant to the problem. This 'brainstorming' extracts a great deal of information in a short time. The factors are classified into checklists which then guide the designers in the collection of relevant data.

Whatever technique is used however, it is at this point that the crucial problems arise. If anything, the improved information that has been collected will have widened the area for search of a solution. The drawback to the conventional design process is a lack of a model of the decisions that need to be taken. The designer tends to go on from here in the abstract and since his answer will depend on his having synthesized a large number of factors each of which has its limitations and constraints, its own set of choices, and its own range of acceptable values, we feel that more could be done to help him through the maze. Because of the complexity we have just described, the practical search method is of necessity carried out by making assumptions about the factors so as to reduce the area of search to a size that is manageable by the single designer or team.

Unfortunately, the factors are nearly always highly interdependent and the whittling-down process that is conducted often leads to dead ends when a complete set of compatible factor choices cannot be found. Designers have tended to fall into the trap of expecting an optimum solution to a total problem to be the sum of optimum solutions to its sub-problems regardless of the fact that the sub-problems are highly interdependent.

J. C. Jones has extended his checklist ideas to cover interdependence by creating an interaction matrix showing which factors on the checklist will affect others. Interactions so found are then treated as sub-problems requiring solution. In our opinion this definition and description of interaction between factors does not go far enough, and so we developed our next model.

COPING WITH INTERDEPENDENCE

For any level we defined each of the factors where there was a choice or range of acceptable answers as a *decision area*. In the field of architecture a decision area might be:

(1) the state of the whole or part of a building such as height, position, direction of span or colour;
(2) individual components of a building such as a window, roof cladding or type of door handle.

The definition of the decision areas will, of course, be dependent upon the particular level under study. The majority of decision areas will be directly linked to or related to others, meaning that the decisions can only be made with reference to one another. In fact, if the total problem structure has been recognized correctly, the majority of such links should be concentrated within the levels rather than betwen them. If we represent a decision area by a point, and a relation or link between decision areas by a line, then the resulting picture is a topological graph. The analysis of interconnected decision areas (AIDA) is the technique or tool with which we try to assist the designer in his hypothesis formulation.

The graph is *not* a directed graph, with some decisions preceding others. Any ordering of the taking of decisions should stem from the structure of the graph and not from preconceived ideas of the relative importance of the decisions. The taking of decisions sequentially is a common procedure and, as we shall see in a later case study, this has its own pitfalls. We wish to preserve the notion at this stage that all decisions are effectively taken simultaneously.

The next step in the process is to examine more closely each of the decision areas. Because each represents a factor about which a decision has to be taken, each will consist of a set of choices or *options* which we will define such that one and only the one option will be selected in the course of making a single decision.

Each decision area link is expanded to determine the relevant option links. We define the mapping of the decision areas,

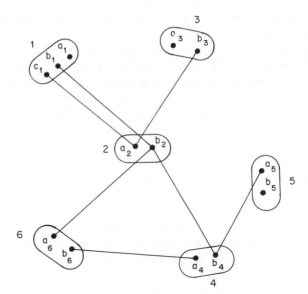

Figure 2
Option graph.

options, and option links as an *option graph*. Figure 2 shows an example of a six decision area option graph, with *a*, *b*, *c* and appropriate subscripts as options within the decision areas.

The option links represent incompatibility* between the options in one decision area and the options in an adjacent decision area. Thus a_1 is compatible with a_2 or b_2 whereas b_1 is not compatible with b_2, and c_1 is not compatible with a_2.

The graphical approach could be used in the assessment of critical sets of decision areas and hence give guidance as to the deployment of the development team. For example, it may be that where a 'ring' exists in the option graph (e.g. Figure 2—decision areas 2, 4 and 6) this is an indicator of a need for a co-ordinated exploratory and problem-solving sub-team. It could be that the option graph for a development project breaks into a number of separated networks, where the implications regarding deployment and control of the team are different again.

Having defined the option graph, we can then find feasible solutions to the problem. A feasible solution includes one option chosen from each area in such a way that there is no contravention of the incompatiblities shown in the option graph. For small problems such as in Figure 2 the solution can be found by simple hand calculations. Larger problems would need to be tackled with the aid of a computer.

Referring to Figure 2, we see that $a_1a_2a_3b_4b_5a_6$ is a feasible solution, while $a_1a_2b_3b_4b_5b_6$ is not. There are in fact only eight

---

* We originally defined an option link as existing where choices in adjacent decision areas were compatible (Harary *et al.*, 1965) but the resulting graphs tended to become heavily congested with lines and were difficult to follow. This newer definition of incompatibility is also more logical since the absence of links between two decision areas implies that they are independent.

feasible solutions in this case and it is interesting to note that none of them includes options c_1, b_2 or b_3.

Where there is a single quantifiable criterion of choice (cost, for example) associated with each option or combination of options, it is possible to find that feasible solution which best satisfies the criterion. It is more likely, however, that compromise will have to be sought between several criteria, which may not be capable of being related to a single scale of measurement.

The descriptions adopted so far have seen the design problem as being deterministic. It is clear that in real life many of the links and criterion values may only be expressible in terms of probabilities. Thus, for example, the feasible link between any two options in adjacent areas may express the fact that some expenditure on research would increase the chance that these options can be made compatible with one another. To handle decisions of this nature, it would be necessary to include statistical decision theory in the methods of choice available within the AIDA technique.

The problem of analysis of option graphs lies principally in their combinatorial nature. There do not need to be many decision areas and options before the analysis must be handled by a computer. Such analysis can be stated as the need to find sets of options, one from each decision area, given the following constraints:

(1) certain pairs of options may be incompatible;
(2) certain higher-order combinations of options may be incompatible;
(3) compatibility in the foregoing cases may be probabilistic rather than deterministic.

THE DESIGN DIALOGUE

The process of design, using AIDA, is foreseen as a dialogue in which working through the logical consequences of the current options and decision areas poses new questions to the design team. The answers to these questions refine the area of search, redefine the decision graphs, define new criteria and combinations of criteria, and make way for the next level of the design process. To facilitate such a dialogue, the following output information may be required at each level:

(1) The number and description of feasible solutions. (If too many, criteria of choice will be required; if too few, new technical options may have to be sought or the initial restrictions and requirements relaxed.)
(2) The number of feasible solutions associated with any specified option or partial set of options.

(3) The number of feasible solutions which meet specified additional criteria and conditions.

(4) The criterion values for all those feasible solutions which are dominant. (A 'dominant' feasible solution is one which meets the condition that there is no other having more favourable values of *all* the criteria.)

(5) A short list of feasible solutions defined, for example, as those which (a) meet specified conditions, and/or (b) head a list of solutions ranked according to one criterion, or (c) are dominant, or (d) have criterion values close to those of the dominant set.

(6) A list of those options which do not appear in any of the feasible solutions contained in the short list.

The number of feasible solutions to a particular problem can be found by treating the decision graph as a directed graph labelled and if necessary augmented by artificial links, so as to have one source and one sink and be acyclic. The option links between each pair of options are written in matrix form with 1 or 0 in the ith row and the jth column representing compatibility or incompatibility between the ith option in the first decision area and the jth option in the second. Artificial links are represented by a matrix of 1's.

The exact number of feasible solutions to the problem can then be calculated by using combinations of matrix multiplication and element-wise matrix multiplication (Hinkley *et al.*, 1967). Manipulation of the matrices can also be made to yield the specification of the options in the solutions and, if a single criterion is available, the value of the best solution.

For larger problems or for more exhaustive analysis of a problem a computer program has been developed. This can provide output information of types (1), (2) and (3) above. Only a single criterion at a time can be dealt with at present.

Returning briefly to our earlier model of the technical content of a level of the design process, we see that AIDA is really a systematic technique for the Synthesis stage. By restructuring the data contained in the information groups into decision areas and options together with their interdependence characteristics, a large number of feasible solutions can be found. In our view the provision of such solutions before Evaluation represents a definite improvement upon the present method of judging each potential solution as it is formulated. It has been our experience in the field of architecture that at best only a few potential solutions are ever explored (admittedly the designer will have chosen some of the better ones in his search process), while, at worst, the first is often accepted because 'it fits the bill' and time is too limited to seek more than marginal improvements.

A further advantage in our approach is in the area of design team co-ordination and individual participation. Again in the field of architecture, consultants brought in to advise on their specialty at an early stage have only a limited view of where their part of the design process fits into the total picture. Design work that proves abortive is often carried out because the implications of another's work are not sufficiently appreciated. If all members of the design team were to assist in the definition of option graphs, these problems of communication might to some extent be overcome. The team's choice of options at any level will give a much tighter and more clear-cut specification within which to plan their detailed design work. Individual members of design teams are too often at different levels of the design process.

AIDA for the most part is a method for synthesis. However, it has been found from experience in trying to apply the method that the analysis stage often benefits from the necessity to provide coherent and orderly information as input material. Similarly, the structure imposed by the AIDA formulation makes it possible to incorporate any available criteria in such a way as to assist in the task of choosing, thus contributing to the stage of evaluation.

A CASE STUDY*

This case study concerns one level of the process of designing a house to be built on a large scale. We were present at many of the design meetings that were held on this project and we were able to observe the emerging design.

The design team developed their designs from conceptualization to the drawing board in a very short period of time. However, in our opinion much of the design team's effort went into exploring, in an intuitive fashion, those aspects of the problem reflected in the linkage between decision areas.

The most important point that we observed in the team's discussions was that they were attempting to make their decisions sequentially, when in fact almost every decision was affected both by those that had gone and those that were yet to come.

Subsequent to their exploration and agreement on a practical solution we drew a graph of the principal decisions and their probable dependence pattern as shown in Figure 4, which relates to the house sketched in Figure 3.

The option graph relating to the house is shown in Figure 5, where for simplicity we have omitted the peripheral decision areas. 'Roof (cladding materials)', 'Drainage' and 'Elevation' together with the two independent decision areas 'Services' and 'Crosswalls'.

* This study arose as part of the Building Industry Communications Research Project carried out jointly by the Tavistock Institute for Human Relations and the Institute for Operational Research.

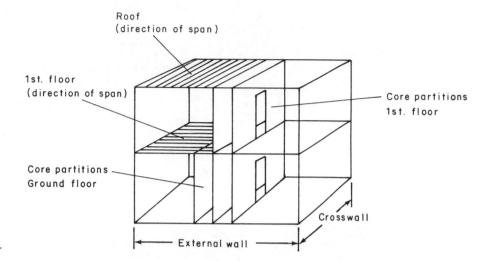

Figure 3
Sketch of house.

Note that the two decision areas in the strategy graph 'Roof (type)' and 'Roof (direction of span)' were linked under the assumption that they were interdependent, whereas the options chosen were subsequently discovered to be fully compatible with each other.

Note also that there are two options within the decision area 'Roof (direction of span)' labelled crosswall. Although these two options perform approximately the same function, they differ both in their compatibility with options in adjacent decision areas and in their cost, and therefore must both be included.

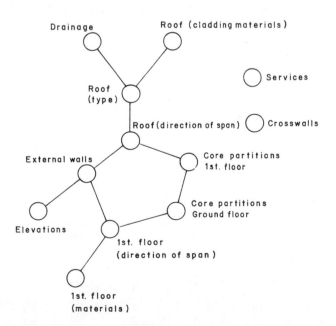

Figure 4
Decision graph for the house.

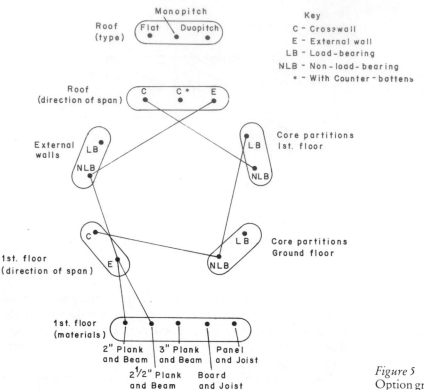

Figure 5
Option graph for the house.

The design team in their discussions (without the benefit of an option graph) had started their exploration with the decision area 'Upper floor (direction of span)'. They made a choice from the options available, and then moved on to discuss other decision areas which were thought to be connected, making

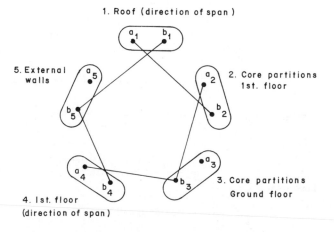

Key: a_1, a_4 – Crosswall a_2, a_3, a_5 – Load-bearing

      b_1, b_4 – External wall b_2, b_3, b_5 – Non load-bearing

Figure 6
Reduced option graph for the house.

their choices and moving on again until they felt they had reached an acceptable total solution.

The main difficulties that the team experienced were recalling their choices of options after more than about two steps, remembering what the decision areas were and whether they were dependent upon one another, and finally reaching options on the opposite side of the circuit that were consistent.

To demonstrate the application of AIDA, we set out to find all the feasible solutions to the problem.

For simplicity of presentation we took up the problem at the point where a duopitch roof had been decided upon. Furthermore, the 'Crosswall (counter battens)' option in the decision area 'Roof (direction of span)' had been ruled out on cost grounds, and the option 'Board and joist' in the decision area 'Upper floor construction (materials)' had already been selected for reasons of cost. The option graph is now reduced to that shown in Figure 6. For ease of calculation we have assigned numbers to the decision areas and letters with subscripts to the options within the decision areas.

It can be shown that there are eight feasible solutions to the problem each of which satisfies all the conditions of compatibility.

The last part of the problem is to choose one of these eight solutions. To obtain a 'best' solution in this example, the design team were pleased to accept any feasible solution provided that it cost as little as possible. We costed the options within each decision area. The cheaper option was given a cost of zero, while the more expensive option was given a cost equal to the difference in cost of the two options. We then compared the total costs of eight feasible solutions. The solutions and their comparative costs are given in Table 1.

Table 1.
Feasible solutions for the circuit of five decision areas

Decision area	Option	Cost	Feasible combinations							
			1	2	3	4	5	6	7	8
1. Roof	a_1 Crosswall span	27	1	1			1			
	b_1 External wall span	0			1	1		1	1	1
2. Core partitions, first floor	a_2 Load-bearing	6	1	1	1			1	1	
	b_2 Non-load-bearing	0				1			1	1
3. Core partitions, ground floor	a_3 Load-bearing	11	1	1	1	1	1	1	1	
	b_3 Non-load-bearing	0								1
4. Upper floor construction	a_4 Crosswall span	0	1	1	1	1				
	b_4 External wall span	13					1	1	1	1
5. External walls	a_5 Load-bearing	23	1		1	1	1	1	1	1
	b_5 Non-load-bearing	0		1						
Total cost of combinations			67	44	40	34	80	53	47	36

The minimum cost solution was in fact the same solution that
the design group had arrived at independently.

THE WIDER VIEW

The decision graphs and option graphs illustrated so far have
been specific to each case and have concerned only part of the
decision process. There is no reason in principle, however, why
a comprehensive map of the total set of important decisions to
be taken at all levels of a project should not be produced. The
resulting picture would be an involved one and its compilation
would be a formidable task. This, however, is no bar to using
the concept of a complete AIDA network as a background
against which such problems as the definition of the roles of
design team members can be discussed.

Once the complete option graph has been compiled,
moreover, it has relevance to all the possible members of a
whole class of designs. There may be areas of decision which are
special to individual cases but the same broad 'shape' of AIDA
networks may well be found to apply over a range of similar
projects. If this turns out to be so, and it would require a good
deal of research to establish the point, then several interesting
possibilities are opened up:

(1) Basic networks of wide usefulness could be prepared and
    made available as a design management aid. At the very
    least this would provide a form of 'checklist' for the
    designers and planners of a new project. This does *not*
    mean, however, that all new designs must look alike. The
    criteria applied to evaluating possible solutions will vary
    widely from case to case.
(2) The factors involved in specific designs could be recorded
    in a manner which would form a useful 'bank' of data and
    experience. This would help to minimize repetition of
    mistakes. It could also help to increase the speed of
    innovation.
(3) Analysis of the structure of the basic network could suggest
    the forms of organization and communications best suited
    to particular classes of design project.

The foregoing possibilities do not only describe the design
process. In particular, (3) above includes organization and
communication and we feel that our earlier models, particularly AIDA, will enable us to understand the structural aspects of
organizations. By studying the interdependence of decisions
that are taken within the whole or part of an organization, we
hope to explore its relevance in deciding the appropriate role
structure of the decision-makers.

In our current work on decision-making in local government
we have been studying city planning. Our models resemble

decision and option graphs, although very much more complex. The decision-makers in these models are the committees, departments, and sections, etc., with certain spheres of responsibilities. The breakthrough will come if we can improve the relationship between the needs of the city and the structure of the organization that has been elected to provide for them.

REFERENCES

Alexander, C. (1964), *Notes on the Synthesis of Form*, Harvard University Press, Cambridge, Mass.

Asimow, M. (1962), *Introduction to Design*, Prentice-Hall, Englewood Cliffs, New Jersey.

Fielden, G. B. R. *et al.* (1963), *Engineering Design*, HMSO, London.

Harary, F., Jessop, W. N., Stringer, J. and Luckman, J. (1965), 'An algorithm for project development', *Nature*, **206**, 118.

Hinkley, D., Jessop, W. N., and Luckman, J. (1967), Unpublished paper IOR/35, Institute for Operational Research.

Jones, J. C. (1966), 'Design methods compared', *Design*, August and September.

Acknowledgements

This paper is an individual team member's presentation of work in which a number of colleagues have taken part, notably Mr W. N. Jessop, Mr J. R. Morgan and Mr J. Stringer.

Part Two

The Structure of Design Problems

Introduction

A second principal area of study for design methodologists has been to understand the special nature of design problems, and to describe their particular structure. If the nature and structure of design problems is understood, then methods for tackling them may be developed with more certainty of success. Of course, the papers already included in Part One made many assumptions about the nature and structure of design problems—that they consist of many interacting factors, that they can be decomposed hierarchically into sub-problems, and so on. In particular, Luckman's 'decision graphs' clearly express certain aspects of problem structure, such as the way decision areas, or sub-problems, are often interlinked in a cyclical manner and thus require comprehensive resolution rather than piecemeal, step-by-step approaches.

Very similar problem structures were encountered by Levin, in his study of 'Decision-making in urban design'. He begins by characterizing the 'system properties' of a town in terms of causes and effects. The 'controllable causes' constitute the design itself, which produces 'controllable effects'; but there are 'uncontrollable' causes and effects, too—i.e. those outside the scope of the urban design process. The designer's task therefore becomes defined as 'to choose the controllable causes and to adjust them in such a way that, under the circumstances defined by the uncontrollable causes, desired controllable effects are obtained'. He goes on to define 'design parameters' as the measures of the 'controllable causes' (i.e. the design), 'dependent variables' as the measures of 'controllable effects', and 'independent variables' as the measures of 'uncontrollable' causes and effects. From this analysis he develops a list of eleven operations which constitute the urban design process.

In discussing how these operations work in practice, Levin reveals several interesting aspects of the structure of design problems. For example, the relationships between the population to be accommodated, the population density, and the required area imply a 'tree-like' decision path—i.e. separate 'branches' (or 'roots') leading to a common 'trunk'. However, Levin found it difficult to identify any clear hierarchy of decisions; the designer 'exercises discretion' in choosing where to start on the decision path. It is also clear that there occur in practice many feedback loops, or cycles of decisions from which it is difficult to escape. Levin gives an example of the cyclical decision structure involved in locating public open space, industry, and housing within the designated area of a new town. This problem structure of 'trees' and 'loops' in urban design is very similar to the decision graph structure identified by Luckman in house design.

Given the pernicious structure of the problem, and the 'information void' to be filled, how does the designer cope? Levin suggests that

A very similar situation arises with a certain class of mathematical riddle, in which there always seems to be insufficient information to give an answer. Then you remember that you do have another bit of information, namely that the problem is to be solvable, and this is the extra information that you need. Likewise the designer knows (consciously or unconsciously) that some ingredient must be added to the information that he already has in order that he may arrive at a unique solution. This knowledge is in itself not enough in design problems, of course. He has to look for the extra ingredient, and he uses his powers of conjecture and original thought to do so.

In many cases the necessary extra ingredient provided by the designer is an 'ordering principle'—hence the frequent emphasis (in most other design fields as well as urban design) on geometrical pattern. So, for example, a town might be designed as a star shape, or on a regular grid; a house might be designed as a cube; a coffee-cup might be designed as a cylinder.

It is geometrical arbitrariness in design which Alexander found so unsatisfactory, and which he seeks to remove from design in his paper with Poyner on 'The atoms of environmental structure'. They object that not only is most building design arbitrary, in that the brief does not define the ultimate geometry, but also that the brief (the 'building programme') is itself arbitrary, in that there is no way of testing it objectively. Most people assume that briefs must be arbitrary; they are a question of subjective values, not of objective facts. But Alexander and Poyner think that this view is wrong: 'We believe that it is possible to define design in such a way that the rightness or wrongness of a building is clearly a question of fact, not a question of value.' Their concern, therefore, is to show that it is possible to write an objectively correct programme, which also directly yields the geometry of a building.

They begin this task by rejecting the earlier methodologists' attempts to clarify and define user requirements or 'needs' as the basis of designing. According to Alexander and Poyner, 'needs' are impossible to ascertain. They substitute instead the concept of 'tendencies'—i.e. what people try to do, when given the opportunity. Thus, 'A tendency is an operational version of a need', and it can be tested objectively by observation of people's behaviour. A statement of a tendency is like a hypothesis, which can be tested, refined, and made more accurate, or else shown to be wrong. As well as tested hypotheses, Alexander and Poyner will accept 'speculative' statements of tendencies, 'Provided that they are stated clearly, so that they can be shown wrong by someone willing to undertake the necessary experiments.'

According to this view there is no design problem until tendencies come into conflict. The problem of environmental design is to arrange the environment so that there are no conflicts. So the design problem consists of the set of conflicts between tendencies which might possibly occur in the particular environment under consideration. This problem is resolved by a matching set of 'relations', which are the geometrical arrangements of the environment which prevent the conflicts occurring.

The designer's task, therefore, is to identify the conflicts of tendencies, then to identify known, or to invent new, relations which prevent those conflicts. To invent a new relation, one starts by identifying aspects of an arrangement which *cause* the conflict, and then tries to find ways of removing those causes. The class of specific examples of non-conflict arrangements will be very large—perhaps infinite, in theory. Therefore the task becomes to define an appropriate relation as an abstract geometric property which captures the essential features of those arrangements which prevent a given conflict of tendencies.

What Alexander and Poyner are trying to do is to establish an externalized, objective body of design knowledge, similar to the body of scientific knowledge. By following their procedure they suggest that design can become 'a cumulative scientific effort', on the basis of defining and improving the body of known design relations. Designing will be objective, not intuitive, and in response to a stated relation a designer will have a clear choice: 'either he must accept it, or he must show that there is a flaw in one of the hypotheses'. He cannot reject the relation merely because he does not like it, and if it is relevant to the environment he is designing then it must be included in his design. Design will therefore become a matter of fact, not of value.

For Alexander and Poyner,

all values can be replaced by one basic value: everything desirable in life can be described in terms of the freedom of people's underlying tendencies. Anything undesirable in life—whether social, economic, or psychological—can always be described as an unresolved conflict between underlying tendencies. Life can fulfil itself only when people's tendencies are running free.

However, the view that not only 'tendencies' conflict but also values conflict in many areas—including environmental design and urban planning—is adopted by Rittel and Webber in their discussion of 'Wicked Problems'. Their paper on 'Dilemmas in a General Theory of Planning', from which the section here is extracted, addressed the question of why so many attempts at large-scale planning fail and result in the public attacking the 'social professions'—i.e. town planners, transport engineers, social planners, etc. They argue that any search for scientific bases for solving problems of social policy is bound to fail, because of the very nature of these problems. Science has been developed to deal with 'tame' problems, whereas planning problems are 'wicked' problems.

Problems in science or engineering have to be 'tame' ones before they are amenable to the conventional problem-solving processes. In particular they have clear goals, and clear criteria for testing their solution.

Wicked problems, in contrast, have neither of these clarifying traits; and they include nearly all public policy issues—whether the question concerns the location of a freeway, the adjustment of a tax rate, the modification of school curricula, or the confrontation of crime.

Rittel and Webber go on to outline ten properties of wicked problems. For example, the first property is that 'There is no definitive formulation of a wicked problem.' For a tame problem, all the necessary information for solving it can be stated, but this is not so for wicked problems, in which the identification of relevant information depends first on the kind of solution being proposed or considered. Stating the problem *is* the problem. 'One cannot understand the problem without knowing about its context; one cannot meaningfully search for information without the orientation of a solution concept; one cannot first understand, then solve.' This therefore leads to a criticism of the early 'systems approach' methods of planning, which relied on exhaustive information collection followed by data analysis and then solution synthesis or the 'creative leap'. Such methods they regard as 'first-generation' ones, which need to be replaced with a 'second generation'. These latter methods 'should be based on a model of planning as an argumentative process in the course of which an image of the problem and of the solution emerges gradually among the participants, as a product of incessant judgment, subjected to critical argument'. (Rittel expands on this view of 'second-generation' planning and design methods in Chapter 5.2.)

The concept of planning and design problems as 'wicked' problems, together with the idea that 'first-generation' design methods were too rigid in their procedures and based on too simple an interpretation of the structure of design problems became quite widely accepted. However, a strong alternative view is presented by Simon in his paper on 'The structure of ill-structured problems'. He argues that there is no clear boundary between 'well-structured' and 'ill-structured' problems, which, in Rittel and Webber's terms, might be interpreted as there being no real distinction between 'tame' and 'wicked' problems. Simon shows that some apparently well-structured problems are not always so, and gives examples of theorem-proving and chess-playing, which Rittel and Webber classified as 'tame' problems.

'Taming' problems so that they can be tackled by conventional problem-solving methods essentially means interpreting or structuring the problems. Simon agrees that, 'In general, the problems presented to problem-solvers by the world are best regarded as ill-structured problems. They become well-structured problems only in the process of being prepared for the problem-solvers.' However, he seems unlikely to agree that any ill-structured problems should be allowed to remain so. His aim is to suggest that conventional problem-solving processes (in particular those familiar in the world of artificial intelligence) can successfully tackle ill-structured problems. Hence his concern to show that some well-structured problems where artificial intelligence procedures have been applied (e.g. theorem-proving and chess-playing) are not as 'well-structured' as may be thought; rather than a clear boundary between well-structured and ill-structured problems, there is a spectrum from one pole to the other.

Simon considers at length an interesting example of how an ill-structured problem may be tackled by conventional procedures—an example of designing a house. He characterizes the problem as ill-structured in terms somewhat similar to those with which Rittel and Webber characterize wicked problems; there is no definite criterion to test a solution, no clearly defined problem space, and so on. To show 'what an architect does' in tackling such a problem, Simon draws on an analysis of the 'thinking-aloud protocol' of (i.e. series of thoughts reported by) a composer writing a fugue, and considers the two processes to have much in common.

According to Simon, the architect begins with 'global specifications' and general attributes of the house, and has, from experience, 'some over-all organization, or executive programme, for the design process itself'. This process is essentially one of hierarchical decomposition: 'The whole design, then, begins to acquire structure by being decomposed into various problems of component design ...'. For Simon's view the important consideration is that, 'During any given short period of time, the architect will find himself working on a problem which, perhaps beginning in an ill-structured state, soon converts itself through evocation from memory into a well-structured problem'. There are some difficulties which may arise from this process. For example, some interrelations among sub-problems may be neglected; some sub-solutions may be found inappropriate at later stages; some criteria may be overlooked. However, these difficulties should be avoided by the architect's skill in his over-all organization of the design process.

Simon goes on to describe this design process in general terms; it comprises a 'general problem-solver' which is capable of working on well-structured sub-problems, plus a retrieval or recognition system which modifies the problem space from time to time, by recognizing the current external state of the problem and evoking from memory appropriate modifications or substitutions to the problem space. Thus the general process alternates between problem-solving in a localized, well-structured problem space and modification of the problem space—in other words, moving from one sub-problem to the next. Simon considers this to be an essential general process because of the serial nature of the problem-solving system and because the problem-solver can only cope with limited inputs and produce limited outputs. This, he suggests, is a more reasonable process that the alternative of bringing all information together at the outset and providing one well-structured problem space.

Simon supports his view of the design process with an account of ship design in the 1920s by Sir Oswyn Murray. The process is described as being organized hierarchically. Nevertheless, there is a crucial stage towards the end of the process which does not seem to be accounted for in Simon's general process. After all the sub-problems have been solved more or less independently, it generally becomes apparent that there are a lot of conflicts and overlaps to be resolved into a complete final design. How is this resolution achieved by the hierarchical problem-solving system? According to Sir Oswyn, 'These difficulties are cleared up by discussion at round-table conferences, where the compromises which will least impair the value of the ship are agreed upon.' This is probably not as easy as it sounds.

A hierarchical decomposition process is normal wherever design processes are necessarily formalized and externalized, Simon claims. Such a process offers a sequence of well-structured problems to a problem-solver of limited capabilities. He therefore concludes that there is no reason to suppose that radically different problem-solving abilities or techniques are needed to tackle ill-structured problems. The main distinction between ill-structured and well-structured problems may be nothing more than the respective sizes of knowledge base.

Clearly there are some significant differences of view between the authors included here on the implications of problem structure for how design problems are, could, and should be tackled. The one point of agreement is that design problems are inherently ill-defined, and the four contributions—from Levin, Alexander and Poyner, Rittel and Webber, and Simon—have offered different views about how to cope with ill-defined problems. According to Levin the designer copes by adding an 'extra ingredient', such as an ordering principle. Alexander and Poyner attempt to eliminate such arbitrariness by devising an objective, externalized body of design knowledge. For Rittel and Webber it is morally objectionable to turn wicked problems into tame ones, and they suggest the need for a participatory 'argumentative' design process. Simon insists that we already have adequate logical procedures for coping even with apparently ill-structured problems.

We are therefore left with a clash of views between those who want to develop an objective 'design science' and those who want to reconstitute the design process in recognition of the ill-defined, wicked, or ill-structured nature of design problems.

FURTHER READING

A pioneering work of design methodology, which found a tree-like structure to engineering design problems, was Marples, D. (1960), *The Decisions of Engineering Design*, Institute of Engineering Designers, London.

Tree-like structures as the basis of urban design are criticized in Alexander, C. (1966), 'A city is not a tree', *Design*, **206**, 46–55.

The approach advocated by Alexander and Poyner is criticized in Daley, J. (1969), 'A philosophical critique of behaviourism in architectural design', in Broadbent, G. and Ward, A. (eds), *Design Methods in Architecture*, Lund Humphries, London; and in March, L. J. (1976), 'The logic of design and the question of value', in March, L. J. (ed.), *The Architecture of Form*, Cambridge University Press, Cambridge.

Alexander subsequently developed his work on 'patterns', most extensively in Alexander, C., Ishikawa, S., and Silverstein, M. (1977), *A Pattern Language*, Oxford University Press, New York.

With reference to the paper by Levin, which reported work carried out at the Building Research Establishment in the mid-sixties, BRE wishes to record the following: 'A recent renewal of BRE research in the context of architectural design has focussed on designers' use of information and the effective application of other research results in building design problems. Several papers stemming from this work have been published, e.g. Information Paper 11/82 which is available, free, from BRE.'

2.1 Decision-making in Urban Design

Peter H. Levin

In the title of this paper I have used the term 'urban design' rather than 'town planning' to emphasize that I am dealing with the essentially innovatory process of environmental design on a large scale, and not with the essentially restrictive and usually piecemeal process of administering town planning controls. To be consistent I shall refer to the practitioners of urban design as urban designers, although they do invariably call themselves town planners.

I have been making a study of the urban design process at fairly close quarters (Levin, 1966), and I set out at the beginning to look for decisions. I still feel that this was a laudable intention, but the execution proved to be exceedingly difficult. It was like trying to put your finger on a blob of mercury: you can't. It's always somewhere else. In just the same way what one thinks is a decision keeps turning out to be merely the consequence of other decisions taken earlier. Sometimes, of course, the converse happens. One can actually witness a decision being made, a specific course of action being decided upon, perhaps at a special meeting called for the purpose. The snag is that a month or two later, say, it becomes apparent that some quite different course of action is being followed.

This situation became so difficult and puzzling that in despair I sat down and wrote the project up in a 12,000-word report without using the word 'decision' once. But that was shirking the problem. Today I want to tackle it.

First of all, what is a decision? I want to adopt a definition that was set out by Professor Elliott Jaques in a talk entitled

Originally published in Building Research Station Note EN 51/66 (1966). Contributed by courtesy of the Director, Building Research Establishment, and reproduced by permission of the Controller, HMSO; Crown copyright, UK.

'The nature of decision-making', given to the Operational Research Society on 15 February 1966. It is perhaps not so much a definition, more a set of characteristics. Professor Jaques sees a decision as a psychological event characterized by:

(1) the exercise of discretion (e.g. in selecting a course of action);
(2) prescribed non-discretionary limits (only within these limits can discretion be exercised);
(3) a goal (towards which the decision-maker is aiming);
(4) committal (i.e. an external event will result from a decision, a wrong decision causing waste or harm in some other form).

The question now is: 'Are there events in the urban design process that possess these characteristics and can, therefore, be called decisions?' If so, what are the operations involved in making such decisions?

On the face of it, urban design is a very much a decision-making process. Set the same problem to a number of designers and they would come up with as many different plans, which suggests that there is discretionary content. They will invariably explain (or complain about) how their solution has been affected by the limits that somebody or something has prescribed, which suggests that these exist too. They will all tell you that they are aiming at something—a goal, in fact, and there can be no doubt that committal is involved. With this encouragement, then, let us look more closely at the design process.

First of all though, let me introduce some more terms. I want to regard a town or a city as a system, a highly complicated system, a vast agglomeration of individuals and groups of people—with a multitude of different motivations and performing a multitude of different activities—enmeshed in a social and economic environment and interacting with buildings and groups of buildings and other artefacts that are located in a specific terrain and subjected to varying climatic conditions.

Many of the properties of this system are measurable, and these measurable properties are essentially effects (in the sense of cause and effect). Some are measured relatively easily: these include the number of people living in a given town at a given time, the mean weekly income per household, and the number of road accidents within a given area in a given time. In addition there is the less easily measured whole host of subjective reactions of each of a whole host of people. For our purposes these effects can be divided into two categories: those which depend in some way on the design—the physical layout—of the town, and those which to all intents and purposes to not depend significantly on the design.

In distinguishing between effects in this way we are also distinguishing between causes. On the one hand the design of the built environment; on the other, all other causes. The design, which is (let us say) within the designer's control, I shall take as constituting a set of 'controllable causes'. Among the other causes are economic and social forces and individual motivations, as well as the weather and the properties of the materials chosen. All these are beyond the designer's control, and I shall call them 'uncontrollable causes'. Those effects which are independent of the design will be called 'uncontrollable effects', and those which do depend on the design (as well as on the uncontrollable causes) will be called 'controllable effects'.

These relationships between causes and effects are shown schematically in Figure 1. This diagram does oversimplify the situation in that it makes no allowance for the fact that effects

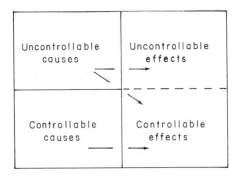

Figure 1

can produce secondary effects, e.g. high mean weekly income per household can result in high car ownership. High income is not a root cause, however, and must therefore be classified as an effect. It should also be remembered that, strictly speaking, the only evidence of a cause is in the effects that it produces. The boundary between uncontrollable and controllable effects is shown blurred, because it is impossible to say quite when the influence of the design changes from being insignificant to being significant.

Using these terms we can now formulate a fairly neat and concise definition of the designer's task. It is to choose the controllable causes and to adjust them in such a way that, under the circumstances defined by the uncontrollable causes, desired controllable effects are obtained. These desired controllable effects, of course, constitute his goal, while it is in choosing and adjusting the controllable causes that he exercises discretion. (It should perhaps be added that a designer is often in the position of choosing, to some extent, his own goals, and in doing this he is again exercising discretion.)

It will be more convenient from now on to talk of measures
of the system, which is what a designer does when he uses terms
such as distance and population density. Let me then define:

(1) design parameters as measures of the controllable causes,
    the design;
(2) independent variables as measures of the uncontrollable
    causes and effects;
(3) dependent variables as measures of the controllable effects.

It may be easier to accept this terminology if one thinks of the
general equation for a straight line, $y = mx + c$. For a given line,
say $y = 3x + 2$, the parameters m and c take specific values (just
as design parameters take fixed specific values when the design
is realized) and the equation then tells us how y (the dependent
variable) varies according to the values taken by x (the
independent variable). At the design stage, of course, the
designer's task is to choose values for m and c such that, for
predicted values of x, y will take the values that he desires.

If we are to study the decisions that the designer makes, we
clearly need more details of how he sets about his task. In
particular, we want to know what operations he carries out in
solving design problems.

It is already evident that the designer must be aware of the
existence of design parameters and independent and dependent
variables. So we already know about three operations:

(1) *The identification of design parameters*
(2) *The identification of independent variables*
(3) *The identification of dependent variables*

It is evident too that he must be aware of cause-and-effect
relationships, while there are other relationships involving
parameters only, or parameters and dependent variables only,
that are in common usage (as shown when areas are added or
subtracted, and when the ratio of two dependent variables—
such as density and population—is a parameter). Hence we
have:

(4) *The identification of relationships among parameters and
    variables*

We also know that the designer needs to *predict* the values that
the independent variables will take, whether directly (e.g. that
the ultimate car ownership in A.D. 2010 will be 0.40 cars per
person) or in terms of the effect produced (e.g. that attitudes to
family size and composition and to overcrowding will be such
as to produce, for a given provision of dwellings, an average of
3.25 people per dwelling). Hence:

(5) *The prediction of values of independent variables*

Now let us look at the dependent variables. These, of course,
are measures of the designer's goal, and to be a goal at all it

must, in some respects at least, be fairly well defined. In practice there are many goals, and they invariably arise out of certain requirements attributed to the clients or to the users.

The designer has, of course, to produce a design that will enable the goals to be attained. The important thing about these requirements is that they are commonly expressed in terms of limits. Thus there will be an upper limit to the amount of money that may be spent or to the population density that will be permitted, and a lower limit to the size of population required to support a theatre or neighbourhood shopping centre. Limitations of this sort I shall refer to as constraints. So we have:

(6) *The identification of constraints governing dependent variables*

These constraints delimit some of the designer's ends rather than limit his means.

We also know that design parameters too are governed by constraints. Some arise if there is an impassable barrier that the development may not cross. Others are derived from user requirements: the rule that the distance from a child's primary school to its home should be less than half a mile is obviously one of these. Hence:

(7) *The identification of constraints governing design parameters*

These constraints are straightforward limits on the means by which the designer's ends may be attained.

We know too that the output of the design process is going to be a design, a diagram on which will be shown parameters with unique values. This is sufficient evidence of:

(8) *The identification of values of design parameters*

It is necessary, moreover, that the designer should know what effects the design is likely to have. This implies another operation:

(9) *The identification of expected values of dependent variables*

A further operation is the result of the fact that a designer cannot take all his information into account at once. He formulates solutions to one or more sub-problems on the basis of some of his information. He then has to find out if these solutions are consistent with each other and with the other information that he has not used yet. In our terms, this is:

(10) *The investigation of the consistency of values, relationships, and constraints*

The final operation in the list arises from the fact that a designer frequently produces several alternative solutions to subproblems or even to the whole problem. He therefore has to compare them and choose between them. The criteria, reasonably enough, are the dependent variables. The values they take in practice, once the design is realized, indicate how successful the design is in coping with the circumstances—that is, the independent variables—actually encountered. This operation, then, is:

(11) *The comparison of, and selection from, alternative sets of values*

Let me now try to show how these operations work out in practice.* For this it will be convenient to use a network representation. I want to take a very simple but real example, which arose in a study for a town expansion.

Consider the following two relationships which the designer had identified:

(1) Population increase P
       =required area A × population density d.
(2) Required area A
       =area A' available within present boundary + area A'' required outside present boundary.

If we represent each variable or parameter by a point, and relationships between them by lines, we have two triangular networks as shown in Figure 2. They are symmetrical, as one would expect, seeing for example that $p = f(A,d)$, $A = f(d,P)$ and $d = f(P,A)$. These particular networks are non-directed, because they do not represent cause-and-effect relationships. An effect is a function of a cause but a cause can hardly be a

---

*It is not uncommon nowadays for papers on the design process to state early on that 'the design process consists of n stages' and to proceed from there. There seems to be the makings here of a contemporary disputation: how many stages does a design process have? Three and four seem to command the greatest popularity, although five, six, and seven have minority support (see for example various contributions in Jones and Thornley, 1963). Each author seems to have used a different terminology, but I think that the most common terms can be plausibly reconciled with these eleven operations.

Briefing and data collection do not, however, correspond to any in my list. I take them to signify the provision of the raw material, the informational context from which may be identified variables and parameters and the constraints upon them.

Problem definition and analysis seem to denote most of the operations of identification.

Identification of form and creation can, I suggest, be equated with the identification of design parameters (and possibly their values). Creation is simply one way of doing this.

Synthesis signifies the putting together of information, identification of solutions by deriving them logically from the identified information. It could also include consistency testing.

Finally *evaluation* must correspond to comparison and selection, and hence to the identification of expected values of dependent variables, since these are the criteria. Consistency testing could come under this heading too.

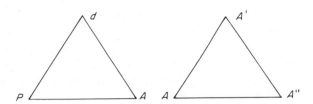

Figure 2

function of an effect, hence the relationship effect $E = f$(cause C) would be represented by a directed line from C to E.

Now, in this study the designer identified on the basis of user requirements values for P and d, and from them derived a value for A. We can use arrows to represent this procedure (noting that they would not oppose any arrows used for the different purpose of denoting cause-and-effect relationships), putting them in on the lines PA and dA. Next this value for A and an already identified value for A' were used to determine A''. This can be represented by the arrows AA'' and $A'A''$. Putting in a dotted line to indicate that A in both triangles is identical, it is clear that the resulting diagram (Figure 3) represents a tree-like

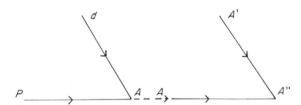

Figure 3

path through the network, and that this path corresponds to that initially taken by the designer. Clearly too the sum total state of the design at any time is given by the collection of parameter values on the tree-like path thus far described.

When the designer had arrived at this value for A'' he tested it against constraints by trying to designate sufficient land on the map. He decided that he could not get it all in satisfactorily, so he designated as much as he could, worked back to find the number of people he could get into it, and took this as the maximum tolerable figure.

The pattern in Figure 3 is strikingly similar to those shown in the pioneering paper by Marples (1960) on *The Decisions of Engineering Design*. There is one important difference, however: Marples found a distinct hierarchy of decisions. There was an immediate problem and in solving it one raised sub-problems. Again, in solving these one raised sub-sub-problems. All the sub-problems and sub-sub-problems had to be solved if the problem was to be solved. In our case—and I think that it is true for the non-engineering aspects of environmental design generally—any hierarchy is much less apparent. There are many immediate problems and they interact more or less on a par with one another. The question then is, where does one start? And the answer is that the designer has discretion. You may remember me enquiring earlier where was the discretionary content in the design process. There is some of it here.

Unfortunately there is also a big procedural problem here too. Let us take another example; a shade hypothetical but realistic and more difficult. It is the problem of where to locate public open space (POS), industry (Ind) and housing (Hsg) within a designated area. These sub-problems are related, and this can be shown on a telescoped network (Figure 4). (In a strict network representation each sub-problem would be shown as a cluster of points, tightly connected internally but also connected to other clusters. In Figure 4 we have simply

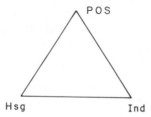

Figure 4

used one point to represent each cluster and a single line to represent all the lines between a pair of clusters.)

You might start by locating first the public open space, not only because your landscape consultant reported first but also because you want to preserve the nicest green spaces, and you suspect too that public open space offers the least room for manoeuvre. Probably you would locate industry next, looking for biggish, flat, ugly sites near likely roads. That done, you come to housing.

Now housing needs to be located in reasonably sized chunks in order to provide adequate catchments for primary schools and neighbourhood facilities. Unfortunately, though, you are likely to find that a fair proportion of the land that remains is in the shape of scruffy little oddments. The ideal solutions for the three sub-problems are inconsistent. In putting the housing right you disarrange your solutions for public open space and industry. When these have in turn been put right, your housing agains needs reforming. You are, in fact, in a cyclic situation. Each problem in turn take priority, and these priorities thus constitute an intransitive set.

Whether one has a cycle of sub-problems or a chain of them (as with *PAA''* in Figure 3), the problem of intransitive priorities is the same. The designer has to balance the solutions to the sub-problems, and in practice does this by circulating round and round a cycle or shuttling to and fro along a chain. Both of these procedures can become very tedious, especially when continual redrawing of plans is involved. At the end of the process, usually when the time limit has expired, there may well be some anomalies remaining. One English new town has two factories in a corner of its town park. Another example of a shuttling process, very common nowadays, is given by the land-use map and the road network. Each is a function of the other, and each alternately takes priority. The situation is rather like a game of ping-pong, in which the designer serves and his traffic consultant returns. The object of the exercise is of course to get a balance, i.e. to get the ball to balance on the top of the net, and one does wonder whether batting it to and fro can be an efficient way of achieving this.

Let me return to the 11 operations of which the design process seems to be constituted. It is clear that once all the information has been generated a recognizable decision-making situation exists. There is a number of goals, specified in terms of dependent variables and constraints upon them. Bearing these goals in mind, the designer has to exercise his discretion in choosing among a number of sets of parameters and parameter values, the area of discretion being limited (by constraints on the parameters). The designer, in making these decisions, is processing certain information. Before he can do this, the information must be generated, and it is noteworthy that information-generating, not less than information-processing, involves making decisions.

We have, then, two categories of operation—information-generating and information-processing. By and large, identification and prediction operations fall into the first category except, for example, where parameters and their values are identified by being derived from other information. Identification by derivation is an information-processing activity, as are consistency testing and comparison/selection. Information-generating may be likened to the extracting of a metal from the ore, and information-processing to the working of a piece of metal in the manufacture of a product.

Now, both the generation and the processing of information involves the making of decisions, and I want to ask: 'How are these decisions made?' How does a designer decide *what* information to feed in next, and *how much* of it, and in *what detail*, and how does he decide *when* to do it? How does a designer decide *what to do* with this information, *when* and *how* to carry out consistency testing or comparison and selection? And, in making these decisions, how much discretion has he? What is it that limits his freedom to exercise this discretion?

Take first the generation of information. Excluding the processing of other information within the design process itself, information sources may be grouped into three main categories:

(1) practical considerations;
(2) documented knowledge and personal experience of user requirements and of previous designs;
(3) conjecture.

These three categories are quite clearly differentiated according to the amount of freedom or discretion that the designer is allowed in selecting information from the source and in implementing it. Practical considerations, which usually limit the designer's means, tend to be well-defined and inflexible. They cannot be ignored and they must be satisfied, hence the designer's discretion is minimal. At the other extreme, a conjecture may be stimulated by quite abstract concepts and may be quite general and ill-defined: the designer is free to produce and modify ideas as he pleases, and in this respect has a maximum of discretion. At the intermediate level of discretion come documented knowledge and personal experience, invariably a mixture of objective description and subjective inference. A designer cannot afford to ignore such information, yet he is free to place his own interpretation on it, to form his own opinion of its validity and relevance to his present problem.

Practical considerations largely have the effect of restricting the shape and form that the design may take, i.e. they are expressed as constraints on design parameters. These considerations stem from

(1) the need to provide utility services (electricity, water, sewerage, etc.);
(2) the need to take into account the nature of the site, present land-uses and existing artefacts;
(3) the need to take into account the constructional methods and materials that are available, and the phasing of the construction; and
(4) the need to obey statutory regulations and government recommendations.

The designer ignores these considerations at his peril.

Documented knowledge and personal experience of user requirements and of previous designs cover a wide field, which the designer must—to begin with—survey. Unfortunately, however, so far as user requirements are concerned it is not so much a field, more a desert, in which facts are exceedingly few and far between. (It is incredible how little objective data there are on the interaction between men and the built environment.) At any rate the designer must survey this field and perceive what information it contains: this information is scrutinized and inferences drawn from it. Both perception and the drawing of inferences act as filtering or selecting processes. Some information may be overlooked or unpalatable facts not faced:

false inferences may be made and alternative inferences not seen. So decisions are made here, involuntary to the extent that the limits are determined by the designer's powers of perception and deduction. (This aspect of decision-making has been discussed, in a somewhat different context, by Abercrombie, 1960.) It may be noted that decisions on user requirements are in the main decisions as to goals: the designer has discretion in formulating the goals of the design as well as in deciding how to attain these goals.

Now, in spite of there being some information in the first two categories, the designer is still left with a very great deal of freedom in choosing design parameters and their values, an informational void. How does he cope with it?

A very similar situation arises with a certain class of mathematical riddle, in which there always seems to be insufficient information to give an answer. Then you remember that you do have another bit of information, namely that the problem is to be solvable, and this is the extra information that you need. Likewise the designer knows (consciously or unconsciously) that some ingredient must be added to the information that he already has in order that he may arrive at a unique solution. This knowledge is in itself not enough in design problems, of course. He has to look for the extra ingredient, and he uses his powers of conjecture and original thought to do so.

Where then is this extra ingredient? In many if not most cases it is an 'ordering principle'. The preoccupation with geometrical patterns that is revealed in many town plans and many writings on the subject demonstrates this very clearly. Zoning according to land-use is also an ordering principle, of course. The philosopher's stone of the urban designer seems to be the little principle with a big effect, that will, like the rule of the road, make order out of chaos. I must confess to being sympathetic to this aim, although I doubt that the answer is likely to come from geometry. After all, shouldn't the aim be to allow for untidiness, wherever possible, rather than to eliminate it?

Clearly, then, there are inherent decisions here, implicit if not explicit, in preferring one principle to another. And the freedom, the discretion to make these choices arises because of the limitation on the information context from which the designer may draw and because of limitations on his means of generating usable information from this context.

So much for information-generating. Let us now look at the processing of information.

There are three distinct categories of information-processing operation:

(1) the derivation of solutions;
(2) consistency testing;
(3) comparison and selection.

The derivation of solutions is used to denote the following of a logical procedure for deriving the solution from the problem. Thus the problem largely defines the solution, just as the measurements of a man define the measurements of his suit. The designer's discretion is obviously limited. Perhaps the closest urban design analogue is the tailoring of a road network to fit expected traffic flows.

It is a noteworthy feature that even so-called 'best' road networks are generally obtained by producing several alternatives and choosing between them, and not by using optimizing procedures. This should not, however, be taken to mean that urban designers are averse to using quantitative data. Far from it. Take traffic. Traffic needs are quantifiable; environmental needs are not. And because quantifiable data are easier to handle, one gets some carefully worked-out plans to deal with the traffic while the environment is more or less left to take care of itself. Of course, arguments backed up by figures always look more convincing than those that are not (even if they depend on inferences and assumptions that one would choose to question). So in the end traffic needs effectively take priority over environmental needs, an illustration of a sort of Gresham's Law: 'Quantitative data drive out qualitative.' This is a good example, I think, of how design procedures and techniques affect the quality of the design that is produced.

Consistency testing, and the comparison and selection of partial solutions, take place repeatedly throughout the design process. This is a reflection of the fact that the solution comes into being in stages, and that not all of the available information is taken into account in formulating partial solutions. Thus derived and conjectured solutions must be checked for consistency with each other and with information gained from knowledge of user needs and practical considerations.

Why does the solution come into being in stages? Clearly because the designer cannot tackle the whole problem in one fell swoop. Evidently what he does is to break up the problem, in such a way—he hopes—as to make solving it easier. Unfortunately, as we have seen, he is liable to end up in a cyclic situation with an intransitive set of priorities, and it may then be very difficult to produce a satisfactory solution. Thus although the sequential treatment of sub-problems may remove some difficulties, it clearly raises others.

In carrying out consistency testing there seems to be little scope for decision-making. That comes in if inconsistencies are discovered. Selection, however, is very much a decision-making operation. Throughout the design process the designer is continually formulating alternative part-solutions: to each sub-problem there may be several such solutions. The total number of possible combinations can be very large indeed. If there are ten alternative solutions to each of ten sub-problems there will be 10^{10} or ten thousand million possible combinations. The designer clearly cannot handle them all, so he thins them out as he goes along.

Unfortunately, when it comes to choosing between solutions there are many criteria by which the comparison can be made, and each solution can thus have a sizeable list of scores. There is evidence from other fields of man's limitations in making judgments in this situation, borne out by—to quote Shepard (1964)—the 'painful hesitation and doubt' with which they are made (especially when contrasted with the 'effortless speed and surety' of, for example, visual perception). Instead, '... there seems to be an overweening tendency to collapse all dimensions into a single "good versus bad" dimension with an attendant loss in detailed information about each configuration or pattern of attributes'.

However it is made, the effect of each selection is of course to limit the field within which later selections can be made. In other words, in making the later decision the designer's discretion is bounded by the prescribed limits imposed by the earlier. If these limits are very narrow the second decision is little more than a consequence of the first.

What, then, can one conclude from this study? It is clear that both information-generating and information-processing involve the making of decisions, unconscious as well as conscious. Moreover, a designer has discretion not only in deciding on means and ends, but also in deciding on the best design procedure, i.e. in deciding how to decide.

In arriving at decisions, the designer's impediments are two-fold. First, there is the lack of objective information as to what ends are desirable and as to the relationships between ends and means. This results in the designer having a very wide area of discretion. Second, there are the designer's own limitations as a human being and the paucity of the tools at his command. Because of these impediments many of his decisions are made with the aim of making the problem easier to solve. In other words, the design that emerges is likely to be the one that was (or appeared to be) the easiest to produce. There is no reason why it should also be the best possible design. In fact, this is unlikely. This is a disquieting situation. How can it be remedied?

First of all, we need a much greater body of knowledge of the interaction between man and the built environment—the hinterland lying between sociology and psychology on the one hand and urban design and architecture on the other. Second, we need techniques by which this knowledge can be sifted and objectively interpreted to yield data relevant to a particular design problem. (For this work some training in deductive reasoning would seem to be required.) Particularly needed are data on the criteria by which designs may be judged. After all, if we have no criteria for comparing mark I and mark II new towns, how can we say that one is any better than the other? In which direction lies progress?

An increase in the amount of relevant factual information should have the effect of curtailing proportionately the designer's area of discretion. Within the area remaining the designer must, of course, be enabled to operate more effectively. Third, then, we need decision-making rules and machinery—techniques for processing data and deriving optimum solutions, techniques that are preferably capable of making use of the products of a designer's innovatory powers as well.

Special pleading for research and development is a commonplace nowadays. There are always deserving causes. But it has been estimated that well before the end of this century we shall have to build more houses, schools, factories, and so forth than we have built in the whole of history. Over £200,000,000,000 (two hundred thousand million pounds) will be spent in doing so. The problems facing urban designers will be of very great magnitude indeed, and they will need correspondingly powerful tools to solve them.

In all that I have said so far, one important question has been begged. To what extent does the functioning of an urban system really depend on the design? As far as social behaviour is concerned, to quote from an article on the subject by Broady (1966): 'there is much to be said for the view that it [design] has, at most, only a marginal effect on social activity'. On the other hand, where amenities such as freedom from traffic dangers or from air pollution are concerned, design can undoubtedly have a major influence.

It may be that the ideal town can best be achieved by a process of refinement, by seeking out the grosser 'mismatches' that appear in existing towns between man and his environment—the acute discomforts, the frustrations, the wastes of human and financial resources—and by simply aiming to alleviate these frictions without introducing new ones.

Alternatively, perhaps it is a valid aim to try deliberately to achieve a situation in which the functioning of the system is as independent as possible of the design. The design, in other words, would have the minimum effect on human organization and activities. What characteristics would such a design have? Clearly it should allow the system to operate in a variety of different ways. It should present each individual with the maximum of freedom, the maximum of choice. This would mean a choice of nearby schools for a child, a choice of convenient shops for a housewife, a choice of nearby friends to call on in an emergency, and a choice of routes to work. This might perhaps involve some overprovision of facilities, but on the other hand local peaks in demand could much more easily be absorbed. Possibly too such a design would be best able to accommodate future unpredicted growth and change in the system.

In ending on this note I fear that I have slipped back into a discussion of ordering principles. Can there be such a thing as an ordering principle that will allow for probabilistic disorder? With greater understanding of the man–environment interaction perhaps we will be able to find the answer.

REFERENCES

Abercrombie, M. L. J. (1960), *The Anatomy of Judgment*, Hutchinson, London.

Broady, M. (1966), 'Social theory in architectural design', *Arena— The Architectural Association Journal*, **81**, (January), 149–154.

Jones, J.C. and Thornley, D. (eds) (1963), *Conference on Design Methods*, Pergamon Press, Oxford.

Levin, P. H. (1966), 'The design process in planning', *Town Planning Review*, **37** (1), 5–20.

Marples, D. L. (1960), *The Decisions of Engineering Design*, Institution of Engineering Designers, London.

Shepard, R. N. (1964), 'On subjectively optimum selection among multiattribute alternatives', in Shelly, M. W., and Bryan, G. L. (eds), *Human Judgments and Optimality*, John Wiley and Sons, New York.

2.2 The Atoms of Environmental Structure

*Christopher Alexander and
Barry Poyner*

The atoms of environmental structure are *relations*. Relations are geometrical patterns. They are the simplest geometrical patterns in a building which can be functionally right or wrong. A list of relations required in a building replaces the design programme (or brief), and the first stages of sketch design.

INTRODUCTION

At present there are two things wrong with design programmes. First of all, even if you state clearly what a building has to do, there is still no way of finding out what the building must be like to do it. The geometry of the building is still a matter for the designer's intuition; the programme does not help with the geometry.

Secondly, even if you state clearly what the building has to do, there is no way of finding if this is what the building *ought* to do. It is possible to make up a very arbitrary programme for a building. There is, at present, no way of being sure that programmes are themselves not arbitrary; there is no way of testing what the programme says.

As far as this second point goes, most designers would maintain that no programme can ever be made non-arbitrary. They would say that the rightness or wrongness of a programme is not a factual matter, but a moral one; it is not a question of

Originally published as part of a Ministry of Public Building and Works Research and Development Paper (1966). Reproduced by permission of the Controller, HMSO.

fact, but a question of value. These people argue in the same way about the physical environment itself. They say that the environment cannot be right or wrong in any objective sense, but that it can only be judged according to criteria, or goals, or policies, or values, which have themselves been arbitrarily chosen.

We believe this point of view is mistaken. We believe that it is possible to define design in such a way that the rightness or wrongness of a building is clearly a question of fact, not a question of value. We also believe that if design is defined in this way, a statement of what a building ought to do, can yield physical conclusions about the geometry of the building, *directly*. We believe, in other words, that is is possible to write a programme which is both objectively correct, and which yields the actual physical geometry of a building.

We shall now describe this new kind of programme. Our argument will have three parts. First, we shall replace the idea of need with its operational counterpart—which we call a *tendency*. Second, we shall show that a single need, when operationally defined, makes no demands on the physical environment—and that the physical environment requires a specific geometry only to resolve *conflicts* between tendencies. Third, we shall show that once a conflict between tendencies is clearly stated, it is then possible to define the geometrical *relation* which is required to prevent the conflict, and to insist that this relation must be present in any building where the conflict can occur. Finally, we assert that the environment needs no geometrical organization, over and above that which it gets from combinations of relations so defined.

1.   WHAT IS A NEED?

Let us begin with the kind of programmes which people write today. It is widely recognized that any serious attempt to make the environment work, must begin with a statement of user needs. Christopher Jones calls them performance specifications; Bruce Archer calls them design goals; in engineering they are often called design criteria; at the Building Research Station they are called user requirements; at the Ministry of Public Building and Works they have been called activities; they are often simply called 'requirements' or 'needs'. Whatever word is used, the main idea is always this: before starting to design a building, the designer must define its purpose in detail. This detailed definition of purpose, goals, requirements, or needs can then be used as a checklist. A proposed design can be evaluated by checking it against the checklist.

But how do we decide that something really is a need? The simplest answer, obviously, is 'Ask the client'. Find out what people need by asking them. But people are notoriously unable to assess their own needs. Suppose then, that we try to assess people's needs by watching them. We still cannot be sure we know what people really need. We cannot decide what is 'really' needed, either by asking questions, or by outside observation, *because the concept of need is not well defined.*

At present the word need has a variety of meanings. When it is said that people need air to breathe, it means that they will die within a few minutes if they don't get it. When someone says 'I need a drink', it means he will feel better after he has had one. When it is said that people 'need' an art museum the meaning is almost wholly obscure. The statement that a person needs something has no well-defined meaning. We cannot decide whether such a statement is true or false.

We shall, therefore, replace the idea of need, by the idea of 'what people are trying to do'. We shall, in effect, accept something as a need if we can show that the people concerned, *when given the opportunity*, actively try to satisfy the need. This implies that every need, if valid, is an active force. We call this active force which underlies the need, a *tendency*.

A tendency, therefore, is an operational version of a need. If someone says that a certain need exists, we cannot test the statement, because we do not know what it really claims. If someone says that a certain tendency exists, we can begin to test the statement.

Here is an example. Suppose we say 'People working in an office need a view'. This is a statement of need. It can be interpreted in many ways. Does it mean 'It would be nice if people in offices had views'? Does it mean 'People will do better work if they have a view'? Does it mean 'People say they want a view from their offices'? Does it mean 'People will pay money to get a view from their offices'? There are so many ways of interpreting it, that the statement is almost useless. We don't know what it really says.

But if we replace it by the statement 'People working in offices try to get a view from their offices', this is a statement of fact. It may be false; it may be true; it can be tested. It is a statement of a tendency. If observation shows that people in an office actively try to get those desks which command a view, it is clearly reasonable to say that they need the view. If, on the other hand, people make no effort to get a view, even when they get the chance, we shall naturally begin to doubt the need.

Now, every statement of a tendency is a hypothesis. It is an attempt to condense a large number of observations by means of a general statement. In this sense a statement of a tendency is like any scientific theory.

Since a statement of a tendency is always a hypothesis—that is, a way of interpreting observations—we must try as hard as possible to rule out alternative hypotheses. Suppose we have observed that people in offices try to get desks near the window when they get the chance. It is possible to infer from this that they are trying to get a view. But we might equally well infer the existence of other tendencies. They could be trying to get more light, or better ventilation, or direct sunlight. Or they may be trying to get something far more complicated; they may want to be in a position, from which they see the light on the faces of their companions—instead of seeing these companions in silhouette against the window.

In order to be confident that people really seek a view, we must make observations which allow us to rule out such alternative interpretations, one by one. For example, suppose we construct an office in which light levels are uniform throughout, because windows are supplemented by artificial light. Do people still try to work near the window in such an office? If they do, we can rule out the possibility that they are merely trying to get more light.

Ruling out all the alternative interpretations we can think of is a laborious and expensive task. Furthermore, in order to make the hypothesis more accurate, we must try to specify just exactly what kind of people seek a view from their offices, during what parts of their work they seek it most, just what aspects of 'view' they are really looking for. ... Again, this is a laborious and expensive task. It is like the task of forming any scientific hypothesis. A good hypothesis cannot be invented overnight; it can be created only by refinement over many years, and by many independent, different observers.

It is therefore vitally important that we do not exaggerate the pseudoscientific aspects of the concept of tendency. Since a tendency is a hypothesis, no tendency can be stated in any absolute or final form. The ideal of perfect objectivity is an illusion—and there is, therefore, no justification for accepting only those tendencies whose existence has been 'objectively demonstrated'. Other tendencies, though they may be speculative, are often more significant from the human point of view. It would be extremely dangerous to ignore such tendencies, just because we have no data to 'support' them. Provided they are stated clearly, so that they can be shown wrong by someone willing to undertake the necessary experiments, it is as important to include these tendencies in the programme as it is to include those tendencies which we are sure about.

2. CONFLICTS

We now face the central problem of design. Given a statement of what people need, how can we find a physical environment which meets those needs?

In order to answer this question we must first define clearly just what we mean by *meeting* needs. This is not as easy as it seems. So long as we are using the word need, the idea of meeting them seems fairly obvious. However, once we replace the idea of need by the idea of tendency, and try to translate the idea of meeting needs in the new language, we shall see that its meaning is not really clear at all.

The idea of needs is passive; but the idea of tendencies is highly active. It emphasizes the fact that, given the opportunity, people will try to satisfy needs for themselves. When we try to interpret the idea of meeting needs in the light of this new emphasis, we see that it is highly ambiguous. To what extent are people expected to meet needs for themselves, and to what extent is the environment expected to do it for them?

Take, for example, a simple situation: a man sitting in a chair. He has various needs. He needs to shift his position every now and then, so as to maintain the circulation in his buttocks and thighs. If he is trying to read, he needs enough light to read by. If he sits in his chair long enough, he will need food for refreshment. He needs ventilation. Under normal circumstances, he is perfectly able to meet those needs for himself. But if we define a good environment as one which meets needs, we should logically be forced to design an environment which meets these needs for him. This conjures up an image of a man lying in an armchair, food being fed to him mechanically, window opening automatically when the room gets stuffy, light being switched on automatically as evening comes, pads in the chair massaging his buttocks to keep the pressure from building up too much in any one place. ...

The image is absurd. It is absurd because the man is perfectly capable of meeting these needs for himself. Indeed, not only is he capable of meeting them for himself, but, for his own well-being, it is almost certain that he should meet them for himself. Man is an adapting organism. A man who is no longer meeting his own needs is no longer adapting. The daily, hourly, process of adapting is the process of life itself; an organism which is no longer adapting is no longer alive.

It is therefore clear that a good environment is not so much one which meets needs, as one which allows men to meet needs for themselves. If we define a need as a tendency, as something which people are trying to do, then we must assume that they will do it whenever they get the chance. The only job which the environment has, is to make sure they get this chance.

Now at first sight it may seem that the argument leads to a dead end. Go back to the example of the man sitting in a chair. Under normal conditions each one of the tendencies which arises in this situation can take care of itself. The man can do everything for himself. There is no problem in the situation. The environment does not require re-design. If needs are defined as tendencies, and if tendencies are capable of taking care of themselves, then why does the environment *ever* require design by designers? Why can't tendencies always be left to take care of themselves? Why can't people be left to adapt to the environment and to shape their own environment as they wish, with the help of bricklayers, carpenters, electricians and so on? If tendencies are active forces, then people will presumably take action whenever the environment is not satisfactory, and meet their own needs for themselves. Why does the environment need design? Why should designers ever take a hand at all?

The answer is this. Under certain conditions, tendencies conflict. In these situations the tendencies cannot take care of themselves, because one is pulling in one direction, and the other is pulling in the opposite direction. Under these kinds of circumstances, the environment *does* need design: it must be re-arranged in such a way that the tendencies no longer conflict.

Let us go back once more to the man sitting in a chair. There are certain chairs, made of canvas slung between wire supports, in which you cannot move about at will, because your body always sinks to the lowest position, and is held there by the canvas. After sitting in one of these chairs for a few minutes you begin to feel uncomfortable; the pressure on certain parts of the body builds up, but you cannot move slightly to reduce this pressure. You try to shift position but you cannot. At first sight it might seem that this is a case where a single tendency simply is not being met; but this is not so. Indeed, the tendency to try to reduce the pressure on your body has a very simple outlet. You can simply get up and walk about. The trouble is, of course, that in many cases there will be another tendency operating, which makes you want to stay sitting where you are (because you are talking to someone, or because you are in the middle of reading something). It is the conflict between your tendency to stay sitting where you are, and your tendency to shift position, which makes a problem. In a properly designed chair this conflict does not occur.

We may therefore replace the simple-minded definition of a good environment as one which meets needs, by the following definition: *A good environment is one in which no two tendencies conflict.*

Of course the conflicts which occur in buildings and cities can be much more complicated than the one we have just described. There can be conflicts between tendencies within a single person, or between tendencies in different people, or between a tendency in one person and a tendency in a group, or between a tendency in a person and some larger tendency which is part of a mass phenomenon. But the principle is always the same. Provided that all the tendencies which occur can operate freely, and are not brought into conflict with other tendencies, the environment in which they are occurring is a good one. It follows then that the environment only requires design in order to prevent conflicts occurring. If we wish to specify the pattern which an environment ought to have, we must begin by identifying all conflicts between tendencies which might possibly occur in the environment.

In summary: until we have managed to see design problems in terms of conflicts between tendencies, there is nothing for the designer to do. So long as we see nothing but isolated tendencies we must assume that they will take care of themselves. We have only succeeded in stating a design problem in a constructive way at that moment when we have stated it as a conflict between tendencies. Since the tendencies in conflict may often be hidden, this is a difficult process which requires a deliberate and inventive search for conflicts.

3. RELATIONS

We design the environment, then, to prevent conflict. We must now start talking about the features of buildings which can help us do this. The features which cause and prevent individual conflicts are not concrete pieces like bricks, or doors, or roofs, or streets; they are geometrical relationships between such concrete pieces. We call these relationships *'relations'*.

Before describing how we invent a new relation, let us look at some examples of well-known relations. Here are five typical relations from a supermarket:

(1) check-out counters are *near* the exit doors;
(2) the stack of baskets and trolleys is *inside* the entrance, and directly *in front* of it;
(3) meat and dairy refrigerators are *at the back* of the store, and all other goods on display are *between* these refrigerators and the check-out counters;
(4) display shelving has a *tapering* cross-section, narrow at the top, and wider at the bottom.
(5) the store is glass-fronted, with aisles running from *front to back, at right angles* to the street.

These relations have become widely copied and typical of supermarkets, because each of them prevents some specific conflict. Here are the conflicts behind each of the five relations:

(1) *Check-out near exit doors.* This relation prevents a conflict between the following tendencies:
    (a) management has to keep all goods on the sales side of the check-outs;
    (b) management is trying to use every square foot of selling space.
(2) *A pile of baskets or trolleys inside the entrance and directly in front of it.* This relation prevents a conflict between the following tendencies:
    (a) management tries to encourage shoppers to use baskets, so that they are not reluctant to pick up extra goods;
    (b) shoppers tend to make as fast as possible for the goods, and are therefore likely to miss the baskets.
(3) *Meat and dairy products at the back of the store, so that all goods are between these counters and the check-outs.* This

relation prevents a conflict between the following tendencies:

(a) management tries to get every shopper to walk past as
many goods as possible;

(b) shoppers visit meat and dairy sections almost every
time they go to the supermarket.

(4) *Display shelving with tapering cross-section, so that even
goods near the ground are clearly visible to shoppers.* This
relation prevents a conflict between the following tendencies:

(a) people tend to walk around a supermarket without
bending down constantly to look for goods;

(b) people want to be able to find the goods they are
looking for without having to ask where they are.

(5) *Glass-fronted supermarket with aisles running back from
the street and at right angles to it.* This relation prevents a
conflict between the following tendencies:

(a) the management is trying to give passers-by a view of
the entire inside of the supermarket, so as to draw them
in;

(b) if the supermarket is on a street most of the passers-by
are walking past the front.

A relation, then, is a geometrical arrangement which prevents a
conflict. No relation can be regarded as necessary to a building,
unless it prevents a conflict which will otherwise occur in that
building. A well-designed building is one which contains
enough relations to prevent any conflicts occurring in it.

So far we have discussed only known relations—those which
exist already. How do we invent a new relation? Obviously, we
start by stating a conflict. But how do we invent a relation
which prevents the conflict? The key fact is this: tendencies are
never inherently in conflict. They are brought into conflict
only by the conditions under which they occur. In order to
solve the conflict, we must invent an arrangement where these
conditions do not obtain.

Here is an example: where a public path turns round the
corner of a building, people often collide. The following
tendencies conflict:

(a) people are trying to see anyone approaching them some
distance ahead, so that they can avoid bumping into them
without slowing down;

(b) going round a corner, people try to take the shortest path.

At a blind corner the first tendency makes people walk well
clear of the corner, the second makes them hug the corner. At a
blind corner the tendencies conflict.

Before we can invent an arrangement which prevents this
conflict, we must find out exactly what makes these tendencies
conflict. It is not possible to invent a geometrical arrangement

which prevents the conflict, until we have identified the aspects of arrangement which *cause* the conflict. In our example there are several aspects of blind corners which we can blame for the conflict: the fact that the corner is solid, the fact that the corner is square, the fact that the ground is unobstructed around the corner.

To eliminate the conflict we must get rid of one or more of these features. If we make the corner transparent, not solid, people will be able to see through it, and can, therefore, see far enough ahead. If we round the corner with a gradual curve, people will be able to see round the corner. If we place a low obstruction at the corner, like a flower tub, people will have to walk around it, and will see each other over the top of it.

It is plain from the example that there are certain arrangements which cause the conflict, and certain 'opposite' arrangements which prevent the conflict. These two classes of arrangement are mutually exclusive. Our task, given any conflict, is to define the class of arrangements which prevents the conflict. This is always difficult. In theory the class is infinite; even in practice it is very large. We must therefore define the class abstractly. We must define an abstract geometric property, shared by all arrangements in the class, and by no others. This is what we mean by a relation. A relation is a precise geometric definition of the class of arrangements which prevent a given conflict. It must be so worded as to *include* all the arrangements that do prevent the conflict, and *exclude* all those which cause it.

Let us continue our example. We have described certain arrangements which cause a conflict at corners, and others which prevent the conflict. Those which prevent it include: a corner made of transparent material, a rounded corner, a tub of flowers so placed that people have to walk clear of the corner. What is the common property which all these good arrangements have, and which the bad arrangements lack? Roughly speaking, the property is this:

If we define a path round the corner at a distance of one foot out from all walls and objects which stick up above the ground, and examine all chords on this path which are less than fifteen feet long, we shall find that none of these chords are, at eye level, obstructed by anything opaque.

This is the relation which prevents the conflict. In a building which contains this relation at all its corners the conflict will not occur.

The conflict in this example happens to be a simple one. However, even when the conflicting tendencies are much larger in scale, or more subtle, the logic is the same. We state the conflict, give examples of arrangements which cause and prevent it, and then try to abstract the relation which defines the latter class.

Two minor points remain. First, conflicting tendencies occur under specific conditions. The relation required to prevent these tendencies conflicting, is only required under these specific conditions. The conditions under which the conflicting tendencies occur must be stated as part of the relation. Thus, the final form of a relation will always be: 'If such and such conditions hold, then the following relation is required.'

Second, the actual process of inventing a relation will not follow the process of finding conflicts and defining relations in strict sequence, as it has been presented here. In practice the statement of tendencies, the statement of conflict, and the statement of the relation, all develop together.

Let us summarize what we have done. We have described a process which has two steps:

(1) identifying a conflict;
(2) deriving a relation from it.

This process for obtaining a relation is objective in the sense that each of its steps is based on a hypothesis which can be tested. The two hypotheses are:

(1) under certain specific conditions such and such potential conflicting tendencies occur;
(2) under these conditions the relation R is both necessary and sufficient to prevent the conflict.

If we cannot show that either of these hypotheses is false, we must then assume that any building where the conflict can occur must contain the relation specified.

In order to create a building in which no tendencies conflict, the designer must try to predict all the conflicts that could possibly occur in it, define the geometric relations which prevent these conflicts, and combine these relations to form a cohesive whole.

4.   THE SCIENTIFIC ATTITUDE TO RELATIONS

The point of view we have presented is impartial. This is its beauty. Because it is impartial it makes possible a sane, constructive, and evolutionary attitude to design. It creates the opportunity for cumulative improvement of design ideas. Everything hinges on one simple question: what does a designer do when faced with a relation which someone else has written?

The traditional point of view about design says that the rightness and wrongness of a relation is a *question of value*. A designer with this point of view will claim that a relation can only be judged by subjectively chosen criteria or values. Since people value things differently we can never be certain that one designer will accept another designer's opinion, and there is therefore no basis for universal agreement. With this point of view the cumulative development of design ideas is impossible.

Our point of view is different. We believe that all values can be replaced by one basic value: everything desirable in life can be described in terms of the freedom of people's underlying tendencies. Anything undesirable in life—whether social, economic, or psychological—can always be described as an unresolved conflict between underlying tendencies. Life can fulfil itself only when people's tendencies are running free. The environment should give free rein to all tendencies; conflicts between people's tendencies must be eliminated.

In terms of this view the rightness or wrongness of a relation is a *question of fact*. Either the relation *does* prevent a conflict between tendencies which occur or it does not.

As we have said, each relation is based on two hypotheses:

(1) that the conflicting tendencies do occur as stated, under the conditions specified;
(2) that the relation proposed is both necessary and sufficient to prevent the conflict between these tendencies.

Faced with a relation stated in this form the designer has two choices: either he must accept it, or he must show that there is a flaw in one of the hypotheses. *Whatever he does he cannot merely reject the relation because he does not like it.*

The body of known relations must therefore grow and improve. Design, if understood as the invention and development of relations, is no longer merely a collection of isolated and disconnected efforts. It becomes a cumulative scientific effort.

2.3 Planning Problems are Wicked Problems

Horst W. J. Rittel and
Melvin M. Webber

A great many barriers keep us from perfecting [an idealized] planning/governing system: theory is inadequate for decent forecasting; our intelligence is insufficient to our tasks; plurality of objectives held by pluralities of politics makes it impossible to pursue unitary aims; and so on. The difficulties attached to rationality are tenacious, and we have so far been unable to get untangled from their web. This is partly because the classical paradigm of science and engineering—the paradigm that has underlain modern professionalism—is not applicable to the problems of open societal systems. One reason the publics have been attacking the social professions, we believe, is that the cognitive and occupational styles of the professions—mimicking the cognitive style of science and the occupational style of engineering—have just not worked on a wide array of social problems. The lay customers are complaining because planners and other professionals have not succeeded in solving the problems they claimed they could solve. We shall want to suggest that the social professions were misled somewhere along the line into assuming they could be applied scientists—that they could solve problems in the ways scientists can solve their sorts of problems. The error has been a serious one.

The kinds of problems that planners deal with—societal problems—are inherently different from the problems that

Originally published as part of 'Dilemmas in a general theory of planning', *Policy Sciences*, **4** (1973), 155–69. Reproduced by permission of Elsevier Scientific Publishing Company.

scientists and perhaps some classes of engineers deal with.
Planning problems are inherently wicked.

As distinguished from problems in the natural sciences,
which are definable and separable and may have solutions that
are findable, the problems of governmental planning—and
especially those of social or policy planning—are ill-defined;
and they rely upon elusive political judgment for resolution.
(Not 'solution'. Social problems are never solved. At best they
are only re-solved—over and over again.) Permit us to draw a
cartoon that will help clarify the distinction we intend.

The problems that scientists and engineers have usually
focused upon are mostly 'tame' or 'benign' ones. As an
example, consider a problem of mathematics, such as solving an
equation; or the task of an organic chemist in analyzing the
structure of some unknown compound; or that of the chess
player attempting to accomplish checkmate in five moves. For
each the mission is clear. It is clear, in turn, whether or not the
problems have been solved.

Wicked problems, in contrast, have neither of these clarifying traits; and they include nearly all public policy issues—
whether the question concerns the location of a freeway, the
adjustment of a tax rate, the modification of school curricula,
or the confrontation of crime.

There are at least ten distinguishing properties of planning-
type problems, i.e. wicked ones, that planners had better be
alert to and which we shall comment upon in turn. As you will
see, we are calling them 'wicked' not because these properties
are themselves ethically deplorable. We use the term 'wicked'
in a meaning akin to that of 'malignant' (in contrast to 'benign')
or 'vicious' (like a circle) or 'tricky' (like a leprechaun) or
'aggressive' (like a lion, in contrast to the docility of a lamb).
We do not mean to personify these properties of social systems
by implying malicious intent. But then, you may agree that it
becomes morally objectionable for the planner to treat a wicked
problem as though it were a tame one, or to tame a wicked
problem prematurely, or to refuse to recognize the inherent
wickedness of social problems.

1.   There is no definitive formulation of a wicked problem

For any given tame problem, an exhaustive formulation can be
stated containing all the information the problem-solver needs
for understanding and solving the problem—provided he
knows his 'art,' of course.

This is not possible with wicked problems. The information
needed to *understand* the problem depends upon one's idea for
solving it. That is to say: in order to *describe* a wicked problem
in sufficient detail, one has to develop an exhaustive inventory
of all conceivable *solutions* ahead of time. The reason is that
every question asking for additional information depends upon

the understanding of the problem—and its resolution—at that time. Problem understanding and problem resolution are concomitant to each other. Therefore, in order to anticipate all questions (in order to anticipate all information required for resolution ahead of time), knowledge of all conceivable solutions is required.

Consider, for example, what would be necessary in identifying the nature of the poverty problem. Does poverty mean low income? Yes, in part. But what are the determinants of low income? Is it deficiency of the national and regional economies or is it deficiencies of cognitive and occupational skills within the labour force? If the latter, the problem statement and the problem 'solution' must encompass the educational process. But, then, where within the educational system does the real problem lie? What then might it mean to 'improve the educational system'? Or does the poverty problem reside in deficient physical and mental health? If so, we must add those etiologies to our information package, and search inside the health services for a plausible cause. Does it include cultural deprivation? spatial dislocation? problems of ego identity? deficient political and social skills?—and so on. If we can formulate the problem by tracing it to some sorts of sources—such that we can say, 'Aha! That's the locus of the difficulty', i.e. those are the root causes of the differences between the 'is' and the 'ought to be' conditions—then we have thereby also formulated a solution. To find the problem is thus the same thing as finding the solution; the problem cannot be defined until the solution has been found.

The formulation of a wicked problem *is* the problem! The process of formulating the problem and of conceiving a solution (or re-solution) are identical, since every specification of the problem is a specification of the direction in which a treatment is considered. Thus, if we recognize deficient mental health services as part of the problem, then—trivially enough—'improvement of mental health services' is a specification of solution. If, as the next step, we declare the lack of community centres one deficiency of the mental health services system, then 'procurement of community centres' is the next specification of solution. If it is inadequate treatment within community centres, then improved therapy training of staff may be the locus of solution, and so on.

This property sheds some light on the usefulness of the famed 'systems approach' for treating wicked problems. The classical systems approach of the military and the space programmes is based on the assumption that a planning project can be organized into distinct phases. Every textbook of systems engineering starts with an enumeration of these phases: 'understand the problems or the mission', 'gather information', 'analyse information', 'synthesize information and wait for the creative leap', 'work out solution', or the like. For wicked problems, however, this type of scheme does not work. One cannot understand the problem without knowing about its context; one cannot meaningfully search for information without the orientation of a solution concept; one cannot first understand, then solve. The systems approach 'of the first generation' is inadequate for dealing with wicked problems. Approaches of the 'second generation' should be based on a model of planning as an argumentative process in the course of which an image of the problem and of the solution emerges gradually among the participants, as a product of incessant judgment, subjected to critical argument. The methods of Operations Research play a prominent role in the systems approach of the first generation; they become operational, however, only *after* the most important decisions have already been made, i.e. after the problem has already been tamed.

Take an optimization model. Here the inputs needed include the definition of the solution space, the system of constraints, and the performance measure as a function of the planning and contextual variables. But setting up and constraining the solution space and constructing the measure of performance is the wicked part of the problem. Very likely it is more essential than the remaining steps of searching for a solution which is optimal relative to the measure of performance and constraint system.

2.   Wicked problems have no stopping rule

In solving a chess problem or a mathematical equation, the problem-solver knows when he has done his job. There are criteria that tell when *the* or *a* solution has been found.

Not so with planning problems. Because (according to Proposition 1) the process of solving the problem is identical with the process of understanding its nature, because there are no criteria for sufficient understanding, and because there are no ends to the causal chains that link interacting open systems, the would-be planner can always try to do better. Some additional investment of effort might increase the chances of finding a better solution.

The planner terminates work on a wicked problem, not for reasons inherent in the 'logic' of the problem. He stops for considerations that are external to the problem: he runs out of time, or money, or patience. He finally says, 'That's good enough', or 'This is the best I can do within the limitations of the project', or 'I like this solution', etc.

3.   Solutions to wicked problems are not true-or-false, but good-or-bad

There are conventionalized criteria for objectively deciding whether the offered solution to an equation, or whether the proposed structural formula of a chemcial compound, is correct or false. They can be independently checked by other qualified persons who are familiar with the established criteria; and the answer will be normally unambiguous.

For wicked planning problems there are no true or false answers. Normally, many parties are equally equipped, interested, and/or entitled to judge the solutions, although none has the power to set formal decision rules to determine correctness. Their judgments are likely to differ widely to accord with their group or personal interests, their special value-sets, and their ideological predilections. Their assessments of proposed solutions are expressed as 'good' or 'bad' or, more likely, as 'better or worse' or 'satisfying' or 'good enough'.

4. There is no immediate and no ultimate test of a solution to a wicked problem

For tame problems one can determine on the spot how good a solution-attempt has been. More accurately, the test of a solution is entirely under the control of the few people who are involved and interested in the problem.

With wicked problems, on the other hand, any solution, after being implemented, will generate waves of consequences over an extended—virtually an unbounded—period of time. Moreover, the next day's consequences of the solution may yield utterly undesirable repercussions which outweigh the intended advantages or the advantages accomplished hitherto. In such cases one would have been better off if the plan had never been carried out.

The full consequences cannot be appraised until the waves of repercussions have completely run out, and we have no way of tracing *all* the waves through *all* the affected lives ahead of time or within a limited time span.

5. Every solution to a wicked problem is a 'one-shot operation'; because there is no opportunity to learn by trial-and-error, every attempt counts significantly

In the sciences, and in fields like mathematics, chess, puzzle-solving, or mechanical engineering design, the problem-solver can try various runs without penalty. Whatever his outcome on these individual experimental runs, it does not matter much to the subject-system or to the course of societal affairs. A lost chess game is seldom consequential for other chess games or for non-chess-players.

With wicked planning problems, however, *every* implemented solution is consequential. It leaves 'traces' that cannot be undone. One cannot build a freeway to see how it works, and then easily correct it after unsatisfactory performance. Large public works are effectively irreversible, and the consequences they generate have long half-lives. Many people's lives will have been irreversibly influenced, and large amounts of money will have been spent—another irreversible act. The same happens with most other large-scale public works and with virtually all public-service programmes. The effects of an experimental curriculum will follow the pupils into their adult lives.

Whenever actions are effectively irreversible and whenever the half-lives of the consequences are long, *every trial counts*. And every attempt to reverse a decision or to correct for the undesired consequences poses another set of wicked problems, which are in turn subject to the same dilemmas.

6.   Wicked problems do not have an enumerable (or an exhaustively describable) set of potential solutions, nor is there a well-described set of permissible operations that may be incorporated into the plan

There are no criteria which enable one to prove that all solutions to a wicked problem have been identified and considered.

It may happen that *no* solution is found, owing to logical inconsistencies in the 'picture' of the problem. (For example, the problem-solver may arrive at a problem description requiring that both A and not-A should happen at the same time.) Or it might result from his failing to develop an idea for solution (which does not mean that someone else might be more successful). But normally, in the pursuit of a wicked planning problem, a host of potential solutions arises; and another host is never thought up. It is then a matter of judgment whethere one should try to enlarge the available set or not. And it is, of course, a matter of judgment which of these solutions should be pursued and implemented.

Chess has a finite set of rules, accounting for all situations that can occur. In mathematics the tool chest of operations is also explicit; so, too, although less rigorously, in chemistry.

But not so in the world of social policy. Which strategies-or-moves are permissible in dealing with crime in the streets, for example, have been enumerated nowhere. 'Anything goes', or at least, any new idea for a planning measure may become a serious candidate for a re-solution: What should we do to reduce street crime? Should we disarm the police, as they do in England, since even criminals are less likely to shoot unarmed men? Or reapeal the laws that define crime, such as those that make the use of marijuana a criminal act, or those that make car theft a criminal act? That would reduce crime by changing definitions. Try moral rearmament and subsitute ethical self-control for police and court control? Shoot all criminals and thus reduce the numbers who commit crime? Give away free loot to would-be thieves, and so reduce the incentive to crime? And so on.

In such fields of ill-defined problems and hence ill-definable solutions, the set of feasible plans of action relies on realistic judgment, the capability to appraise 'exotic' ideas and on the amount of trust and credibility between planner and clientele that will lead to the conclusion, 'OK let's try that'.

7. Every wicked problem is essentially unique

Of course, for any two problems at least one distinguishing property can be found (just as any number of properties can be found which they share in common), and each of them is therefore unique in a trivial sense. But by '*essentially* unique' we mean that, despite long lists of similarities between a current problem and a previous one, there always might be an additional distinguishing property that is of overriding importance. Part of the art of dealing with wicked problems is the art of not knowing too early which type of solution to apply.

There are no *classes* of wicked problems in the sense that principles of solution can be developed to fit *all* members of a class. In mathematics there are rules for classifying families of problems—say, of solving a class of equations—whenever a certain, quite-well-specified set of characteristics matches the problem. There are explicit characteristics of tame problems that define similarities among them, in such fashion that the same set of techniques is likely to be effective on all of them.

Despite seeming similarities among wicked problems, one can never be *certain* that the particulars of a problem do not override its commonalities with other problems already dealt with.

The conditions in a city constructing a subway may look similar to the conditions in San Francisco, say; but planners would be ill-advised to transfer the San Francisco solutions directly. Differences in commuter habits or residential patterns may far outweigh similarities in subway layout, downtown layout, and the rest. In the more complex world of social policy planning, every situation is likely to be one-of-a-kind. If we are right about that, the direct transference of the physical-science and engineering thoughtways into social policy might be dysfunctional, i.e. positively harmful. 'Solutions' might be applied to seemingly familar problems which are quite incompatible with them.

8. Every wicked problem can be considered to be a symptom of another problem

Problems can be described as discrepancies between the state of affairs as it is, and the state as it ought to be. The process of resolving the problem starts with the search for causal explanation of the discrepancy. Removal of that cause poses another problem of which the original problem is a 'symptom'. In turn, it can be considered the symptom of still another, 'higher level' problem. Thus 'crime in the streets' can be considered as a symptom of general moral decay, or permissiveness, or deficient opportunity, or wealth, or poverty, or whatever causal explanation you happen to like best. The level at which a problem is settled depends upon the self-confidence of the analyst and cannot be decided on logical grounds. There is nothing like a natural level of a wicked problem. Of course, the higher the level of a problem's formulation, the broader and more general it becomes: and the more difficult it becomes to do something about it. On the other hand, one should not try to cure symptoms: and therefore one should try to settle the problem on as high a level as possible.

Here lies a difficulty with incrementalism, as well. This doctrine advertises a policy of small steps, in the hope of contributing systematically to overall improvement. If, however, the problem is attacked on too low a level (an increment), then success of resolution may result in making things worse, because it may become more difficult to deal with the higher problems. Marginal improvement does not guarantee overall improvement. For example, computerization of an administrative process may result in reduced cost, ease of operation, etc. But at the same time it becomes more difficult to incur structural changes in the organization, because technical perfection reinforces organizational patterns and normally increases the cost of change. The newly acquired power of the controllers of information may then deter later modifications of their roles.

Under these circumstances it is not surprising that the members of an organization tend to see the problems on a level below their own level. If you ask a police chief what the problems of the police are, he is likely to demand better hardware.

9. The existence of a discrepancy representing a wicked problem can be explained in numerous ways. The choice of explanation determines the nature of the problem's resolution

'Crime in the streets' can be explained by not enough police, by too many criminals, by inadequate laws, too many police, cultural deprivation, deficient opportunity, too many guns, phrenologic aberrations, etc. Each of these offers a direction for attacking crime in the streets. Which one is right? There is no rule or procedure to determine the 'correct' explanation or combination of them. The reason is that in dealing with wicked problems there are several more ways of refuting a hypothesis than there are permissible in the sciences.

The mode of dealing with conflicting evidence that is customary in science is as follows: 'Under conditions C and assuming the validity of hypothesis H, effect E must occur. Now, given C, E does not occur. Consequently H is to be refuted.' In the context of wicked problems, however, further modes are admissible: one can deny that the effect E has not occurred, or one can explain the non-occurence of E by intervening processes without having to abandon H. Here is an example: Assume that somebody chooses to explain crime in the streets by 'not enough police'. This is made the basis of a plan, and the size of the police force is increased. Assume further that in the subsequent years there is an increased number of arrests, but an increase of offences at a rate slightly lower than the increase of GNP. Has the effect E occurred? Has crime in the streets been reduced by increasing the police force? If the answer is no, several non-scientific explanations may be tried in order to rescue the hypothesis H ('Increasing the police force reduces crime in the streets'): 'If we had not increased the number of officers, the increase in crime would have been even greater'; 'This case is an exception from rule H because there was an irregular influx of criminal elements'; 'Time is too short to feel the effects yet'; etc. But also the answer 'Yes E has occurred' can be defended: 'The number of arrests was increased', etc.

In dealing with wicked problems the modes of reasoning used in the argument are much richer than those permissible in the scientific discourse. Because of the essential uniqueness of the problem (see Proposition 7) and lacking opportunity for rigorous experimentation (see Proposition 5), it is not possible to put H to a crucial test.

That is to say, the choice of explanation is arbitrary in the logical sense. In actuality, attitudinal criteria guide the choice. People choose those explanations which are most plausible to them. Somewhat but not much exaggerated, you might say that everybody picks that explanation of a discrepancy which fits his intentions best and which conforms to the action-prospects that are available to him. The analyst's 'world view' is the strongest determining factor in explaining a discrepancy and, therefore, in resolving a wicked problem.

10. The planner has no right to be wrong

As Karl Popper argues in *The Logic of Scientific Discovery*, it is a principle of science that solutions to problems are only hypotheses offered for a refutation. This habit is based on the insight that there are no proofs to hypotheses, only potential refutations. The more a hypothesis withstands numerous attempts at refutation, the better its 'corroboration' is considered to be. Consequently, the scientific community does not blame its members for postulating hypotheses that are later refuted—so long as the author abides by the rules of the game, of course.

In the world of planning and wicked problems no such immunity is tolerated. Here the aim is not to find the truth, but to improve some characteristics of the world where people live. Planners are liable for the consequences of the actions they generate; the effects can matter a great deal to those people that are touched by those actions.

We are thus led to conclude that the problems that planners must deal with are wicked and incorrigible ones, for they defy efforts to delineate their boundaries and to identify their causes, and thus to expose their problematic nature.

2.4 The Structure of Ill-structured Problems

Herbert A. Simon

1. INTRODUCTION

Certain concepts are defined mainly as residuals—in terms of what they are not. Thus a UFO is an aerial phenomenon not explainable in terms of known laws and objects; and ESP is communication between persons, without evidence of the transmission of signals of any kind.

In just the same way, 'ill-structured problem' (ISP) is a residual concept. An ISP is usually defined as a problem whose structure lacks definition in some respect. A problem is an ISP if it is not a WSP (well-structured problem).

Residual categories are tenacious: it is extremely difficult, or even impossible, to prove that they are empty. The scope of a residual category can be narrowed progressively by explaining previously unexplained phenomena; it cannot be extinguished as long as a single phenomenon remains unexplained.

In this paper I wish to discuss the relation between ISPs and WSPs with the aim of asking whether problems regarded as ill-structured are inaccessible to the problem solving systems of artificial intelligence in ways that those regarded as well-structured are not. My aim will not be to restrict the class of problems we regard as ISPs—in fact I shall argue that many kinds of problems often treated as well-structured are better regarded as ill-structured. Instead, I will try to show that there is no real boundary between WSPs and ISPs, and no reason to think that new and hitherto unknown types of problem-solving

Originally published in *Artificial Intelligence*, **4** (1973), 181–200. Reproduced by permission of North-Holland Publishing Company.

processes are needed to enable artificial intelligence systems to solve problems that are ill-structured.

Some years ago, Walter Reitman (1964, 1965) provided the first extensive discussion of ISPs (which he called 'ill-defined problems'). More recently, the topic has been developed in a somewhat different vein by Allen Newell (1969), who emphasized the relation between problem structure and problem-solving methods. Newell characterized the domain of ISPs as the domain in which only weak problem-solving methods were available.

In this account, I continue the discussion begun by Reitman (and on which I earlier made some preliminary remarks in Simon (1972), Section 5, pp. 274–6). I shall try to give a positive characterization of some problem domains that have usually been regarded as ill-structured, rescuing them from their residual status; and then I shall ask whether the methods used in contemporary artificial intelligence systems are adequate for attacking problems in these domains. I shall not prejudge whether the methods applicable to these problems are weak or strong, but shall leave that to be decided after the fact.

The first section sets forth a set of strong requirements that it is sometimes asserted a task must meet in order to qualify as a WSP. Each of these requirements is examined, in order to characterize the kinds of ISPs that fail to satisfy it. The meaning of the requirements, and their relation to the power of the available problem-solving systems, is then explored further by considering some specific examples of WSPs and of ISPs. Finally, this exploration provides the basis for a description of problem-solving systems that are adapted to attacking problems in the domains usually regarded as ill-structured.

2. WELL-STRUCTURED PROBLEMS

For reasons that will become clear as we proceed, it is impossible to construct a formal definition of 'well-structured problem'. Instead, we must be content simply to set forth a list of requirements that have been proposed at one time or another as criteria a problem must satisfy in order to be regarded as well-structured. A further element of indefiniteness and relativity arises from the fact that the criteria are not absolute, but generally express a relation between characteristics of a problem domain, on the one hand, and the characteristics and power of an implicit or explicit problem-solving mechanism, on the other.

With these caveats, we will say that a problem may be regarded as well-structured to the extent that it has some or all of the following characteristics:

(1) There is a definite criterion for testing any proposed solution, and a mechanizable process for applying the criterion.

(2) There is at least one problem space in which can be represented the initial problem state, the goal state, and all other states that may be reached, *or considered*, in the course of attempting a solution to the problem.

(3) Attainable state changes (legal moves) can be represented in a problem space, as transitions from given states to the states directly attainable from them. But considerable moves, whether legal or not, can also be represented—that is, all transitions from one considerable state to another.

(4) Any knowledge that the problem-solver can acquire about the problem can be represented in one or more problem spaces.

(5) If the actual problem involves acting upon the external world, then the definition of state changes and of the effects upon the state of applying any operator reflect with complete accuracy in one or more problem spaces the laws (laws of nature) that govern the external world.

(6) All of these conditions hold in the strong sense that the basic processes postulated require only practicable amounts of computation, and the information postulated is effectively available to the processes—i.e. available with the help of only practicable amounts of search.

As I have warned, these criteria are not entirely definite. Moreover, phrases like 'practicable amounts of computation' are defined only relatively to the computational power (and patience) of a problem-solving system. But this vagueness and relativity simply reflect, as I shall try to show, the continuum of degrees of definiteness between the well-structured and ill-structured ends of the problem spectrum, and the dependence of definiteness upon the power of the problem-solving techniques that are available.

2.1. The general problem solver

If a problem has been formulated in such a way that it can be given to a program like the General Problem Solver (GPS), can we say that it is a WSP? Before GPS can go to work on a problem, it requires:

(1) a description of the solution state, or a test to determine if that state has been reached;

(2) a set of terms for describing and characterizing the initial state, goal state and intermediate states;

(3) a set of operators to change one state into another, together with conditions for the applicability of these operators;

(4) a set of differences, and tests to detect the presence of these differences between pairs of states;

(5) a table of connections associating with each difference one or more operators that is relevant to reducing or removing that difference.

The first three requirements for putting a problem in a form
suitable for GPS correspond closely to the first three characteristics of a WSP. The fourth and fifth requirements for putting a
problem in a form suitable for GPS correspond closely to the
fourth characteristic of a WSP. Since GPS operates on the
formally presented problem, and not on an external real world,
the fifth requirement for a WSP appears irrelevant to GPS.

In our description of the conditions for GPS's applicability,
it is implicit that the sixth requirement for WSPs is also
satisfied, for the operators and tests mentioned above can all
presumably be executed with reasonable amounts of computation. This does *not* imply, of course, that any problem
presented to GPS within the defined domain can be solved with
only reasonable amounts of computation. Many problems may
not be solvable at all. Of those that are solvable in principle,
many may require immense numbers of applications of operators and tests for their solution, so that the total amount of
computation required may be impractical.

Thus, it would appear at first blush that all problems that can
be put in proper form for GPS can be regarded as WSPs. But
what problem domains satisfy these, or similar, requirements?
Let us examine a couple of possible examples.

2.2.   Is theorem-proving a WSP?

Consider what would appear to be an extreme example of a
WSP: discovering the proof of a theorem in formal logic.
Condition 1 for a WSP will be satisfied if we have a mechanical
proof-checker. Condition 2 might be regarded as satisfied by
identifying the problem space with the space of objects that can
be described in terms of wffs. However, we should note that
limiting the problem-solver in this way excludes it from even
considering expressions that are not wffs.

The same reservation must be made with respect to Condition 3: definitions of the axioms, the rules of inference, and the
processes for applying the latter determine the legal moves and
attainable state changes; but the problem-solver may wish to
consider inferences without determining in advance that they
meet all the conditions of 'legality'—e.g., working backwards
from unproved wffs. Hence the set of considerable moves is not
determined uniquely by the set of legal moves.

Satisfying Condition 4 is even more problematic. There is no
difficulty as long as we restrict ourselves to the object language
of the logic under consideration. But we have no reason to
exclude metalinguistic knowledge, knowledge expressed in a
model space, or even analogical or metaphorical knowledge. A
human theorem-prover, using a metalanguage, may prove a
theorem that is not provable in the object language; or may use
a truth table as a model for solving a problem in the
propositional calculus; or may use the proof of one theorem as

an analogical guide to the proof of another that seems, in some respect, to be similar to the first.

Of course there is nothing magical here in the problem-solver being human. Mechanical systems can be, and have been, given the same kinds of capabilities. (For an example of the use of metalinguistic techniques in theorem-proving, see Pitrat, 1966; for the use of analogies, see Kling, 1971.) What some notions of well-structuredness require, however, is that these capabilities be defined in advance, and that we do not allow the problem-solver to introduce new resources that 'occur' to him in the course of his solution efforts. If this condition is imposed, a problem that admits restructuring through the introduction of such new resources would be an ill-structured problem.

A problem that is not solvable with reasonable amounts of computation when all knowledge must be expressed in terms of the original problem space may be easily solvable if the solver is allowed to (or has the wits to) use knowledge in another space. It follows that, under a literal interpretation of Condition 4, problems of discovering proofs in formal logic are not, for all problem-solvers, well structured.*

Condition 5 is always satisfied in theorem-proving since there is no external 'real world' to be concerned about. Condition 6 is usually satisfied, as far as the basic processes are concerned. Nevertheless, because of the enormous spaces to be searched, contemporary mechanical theorem-provers, confronted with difficult theorems, usually fail to find proofs. It is sometimes said that they could only succeed if endowed with ingenuity; but ingenuity, whatever it is, generally requires violating Condition 4—moving out into the broader world of ISPs.

2.3.   Is chess-playing a WSP?

Next to theorem-proving, the world of games would appear to offer the best examples of well-structuredness. All of the reservations, however, that applied to the well-structuredness of theorem-proving apply to game-playing as well. In addition, new reservations arise with respect to Condition 1—the solution criterion—and Condition 5—the correspondence between the inner world of thought and the outer world of action on real chess boards. Let us consider these two matters in more detail.

In both cases the difficulty stems from the immense gap between computability *in principle* and practical computability

---

*Notice that we are not appealing here to formal undecidability or incompleteness. Our concern throughout is with effective or practicable solvability using reasonable amounts of computation. Problems may be (and often are) unsolvable in this practical sense even in domains that are logically complete and decidable.

in problem spaces as large as those of games like chess. In principle, the concept of 'best move' is well defined; but in practice this concept has to be replaced by, say, maximizing some approximate evaluation function. When a chess-playing program has found (if it does) the move that maximizes this function, it can still be far from finding the move that will win the game of chess—as the modest ability of the best contemporary programs testifies.

In terms of Condition 5, it is not hard to define the WSP of playing an approximate kind of 'chess', where 'winning' means maximizing the postulated evaluation function. But the values of moves as calculated by the approximate evaluation function are simply a means for predicting the actual consequences of the moves in the real game 'outside'. Feedback in terms of the expected or unexpected moves of the opponent and the expected or unexpected board situations arising from those moves call for new calculations by the problem-solver to make use of the new information that emerges.

The ill-structuredness, by the usual criteria, of chess playing becomes fully evident when we consider the play of an entire game, and do not confine our view to just a single move. The move in the real game is distinguished from moves in dynamic analysis by its irrevocability—it has real consequences that cannot be undone, and that are frequently different from the consequences that were anticipated. Playing a game of chess— viewing this activity as solving a single problem—involves continually redefining what the problem is. Even if we regard chess playing as a WSP in the small (i.e. during the course of considering a single move), by most criteria it must be regarded as an ISP in the large (i.e. over the course of the game).

2.4.   Summary: the elusiveness of structure

As our two examples show, definiteness of problem structure is largely an illusion that arises when we systematically confound the idealized problem that is presented to an idealized (and unlimitedly powerful) problem-solver with the actual problem that is to be attacked by a problem-solver with limited (even if large) computational capacities. If formal completeness and decidability are rare properties in the world of complex formal systems, effective definability is equally rare in the real world of large problems.

In general, the problems presented to problem-solvers by the world are best regarded as ISPs. They become WSPs only in the process of being prepared for the problem-solvers. It is not exaggerating much to say that there are no WSPs, only ISPs that have been formalized for problem-solvers.

A standard posture in artificial intelligence work, and in theorizing in this field, has been to consider only the idealized problems, and to leave the quality of the approximation, and the processes for formulating that approximation, to informal discussion outside the scopes both of the theory and of the problem-solving programs. This is a defensible strategy, common to many fields of intellectual inquiry; but it encourages allegations that the 'real' problem-solving activity occurs while providing a problem with structure, and not after the problem has been formulated as a WSP. As Newell and I have observed elsewhere (Newell and Simon, 1972, p. 850, footnote 20) these allegations are refuted simply by observing that 'if [they] were correct, and tasks from the same environment were presented sequentially to a subject, only the first of them would present him with a problem, since he would not need to determine a new problem space and program for the subsequent tasks'. Nevertheless, there is merit to the claim that much problem solving effort is directed at structuring problems, and only a fraction of it at solving problems once they are structured.

3.   ILL-STRUCTURED PROBLEMS

Perhaps something is to be learned by turning the question around. We have generally asked how problems can be provided with sufficient structure so that problem-solvers like GPS can go to work on them. We may ask instead how problem-solvers of familiar kinds can go to work even on problems that are, in important respects, ill-structured. Since the problem domains that have been most explored with mechanical techniques fail in several ways to satisfy the requirements for WSPs, perhaps we have exaggerated the essentiality of definite structure for the applicability and efficacy of these techniques. Perhaps the tricks that have worked in relatively well-structured domains can be extended to other domains that lie far over toward the ISP end of the spectrum.

To explore this possibility we will again examine several examples. Each example will illustrate some specific facet (or several facets) of ill-structuredness. Analysis of these facets will provide us with a positive characterization of ISPs, rescuing them from the status of a residual category. With this positive characterization in hand, we will be in a better position to set forth the capabilities a problem-solving system must have in order to be able to attack problems that are initially ill-structured in one or more ways.

3.1.   Designing a house

It will generally be agreed that the work of an architect—in designing a house, say—presents tasks that lie well toward the ill-structured end of the problem continuum. Of course this is only true if the architect is trying to be 'creative'—if he does not begin the task by taking off his shelf one of a set of standard house designs that he keeps there.

The design task (with this proviso) is ill-structured in a number of respects. There is initially no definite criterion to test a proposed solution, much less a mechanizeable process to apply the criterion. The problem space is not defined in any meaningful way, for a definition would have to encompass all kinds of structures the architect might at some point consider (e.g., a geodesic dome, a truss roof, arches, an A-frame, cantilevers, and so on and on), all considerable materials (wood, metal, plexiglass, ice—before you object, I must remind you it's been done—reinforced concrete, camels' hides, field stone, Vermont marble, New Hampshire granite, synthetic rubber, ...), all design processes and organizations of design processes (start with floor plans, start with list of functional needs, start with facade, ...).

The hopelessness of even trying to sketch the congeries of elements that might have to be included in the specification of a problem space proves the greater hopelessness of defining in reasonable compass a problem space that could not, at any time during the problem-solving process, find its boundaries breached by the intrusion of new alternatives. The second, third, and fourth characteristics of a WSP appear, therefore, to be absent from the house design problem.

The fifth characteristic is also lacking. One thing an architect often does is to make renderings or models of the projected structure. He does this partly because these productions predict, more accurately than other means, properties that the real-world structure will possess if it is actually built. Viewing a model, the architect can detect relations among components of the design that were not available to him directly from his plans. Of course, even the renderings and models fall far short of predicting the actual characteristics of the real building, or the way in which the laws of nature will operate upon it and affect it. Hence, while Frank Lloyd Wright was not greatly disturbed by a leaking roof, it can hardly be supposed that he designed his roofs to leak, which they often did. Nor was the action of New York's atmosphere on the surface of the Seagram Building, and its consequent change of colour, predicted. Doors stick, foundations settle, partitions transmit noise, and sometimes even happy accidents (examples of these are harder to come by) conspire to make the building, as it actually exists and is used, something different from the building of the plans.

Finally, even if we were to argue that the problem space can really be defined—since anything the architect thinks of must somehow be generated from, or dredged from, his resources of memory or his reference library—some of this information only shows up in late stages of the design process after large amounts of search; and some of it shows up, when it does, almost accidentally, Hence, the problem is even less well defined when considered from the standpoint of what is actually known at any point in time than when considered from the standpoint of what is knowable, eventually and in principle.

All of this would seem to make designing a house a very different matter from using GPS to solve the missionaries and cannibals puzzle, or from discovering the proof of a theorem in the predicate calculus. It surely is different, but I shall try in the next paragraphs to show that understanding what the architect does goes far toward bridging the gulf between these problem domains. In this I shall be following quite closely the path first blazed by Walter Reitman (1965, ch. 8) in his analysis of the thinking-aloud protocol of a composer writing a fugue. It should not, of course, appear surprising that a house-designing process would have much in common with a process of musical composition. Showing that they do have much in common is critical to the attempt here to provide a positive characterization of the processes for solving ISPs.

3.2. The architect's processes

Reitman uses the term 'constraints' quite broadly to refer to any or all of the elements that enter into a definition of a problem. He observes (1965, p. 169):

One of the interesting features of many of the problem instances ... is that even though they generally would be considered complex, they include very few constraints as given. Composing a fugue is a good example. Here the main initial constraint, and it is an open constraint at that [i.e. one that is incompletely specified], is that the end product be a fugue. All other constraints are in a sense supplementary, generated from one transformation of the problem to the next.

Similarly the architect begins with the sole problem of designing a house. The client has presumably told him something of his needs, in terms of family size or number of rooms, and his budget (which the architect will multiply by 1.5 or 2 before accepting it as a constraint). Additional specification will be obtained from the dialogue between architect and client, but the totality of that dialogue will still leave the design goals quite incompletely specified. The more distinguished the architect, the less expectation that the client should provide the constraints.

Evaluating the specifications

We can imagine a design process that proceeds according to the following general scheme. Taking the initial goals and constraints, the architect begins to derive some global specifications from them—perhaps the square footage or cubic footage of the house among them. But the task itself, 'designing a house', evokes from his long-term memory a list of other attributes that will have to be specified at an early stage of the design: characteristics of the lot on which the house is to be built, its general style, whether it is to be on a single level or multi-storied, type of frame, types of structural materials and of sheathing materials, and so on. The task will also evoke from memory some over-all organization, or executive programme, for the design process itself.

Neither the guiding organization nor the attributes evoked from memory need at any time during the process provide a complete procedure nor complete information for designing a house. As a matter of fact, the entire procedure could conceivably be organized as a system of productions, in which the elements already evoked from memory and the aspects of the design already arrived at up to any given point, would serve as the stimuli to evoke the next set of elements.

Whether organized as a system of productions or as a system of sub-routine calls, the evocation of relevant information and sub-goals from long-term memory can be sequential. As Reitman says of the fugue (1965, p. 169):

Just as 'sentence' transforms to 'subject plus predicate', and 'subject' may transform to 'article plus noun phrase' ..., so 'fugue' may be thought of as transforming to 'exposition plus development plus conclusion', 'exposition' to 'thematic material plus countermaterial', and 'thematic material' to 'motive plus development of motive'.

Applying the same linguistic metaphor to house design, 'house' might transform to 'general floor plan plus structure', 'structure' to 'support plus roofing plus sheathing plus utilities', 'utilities' to 'plumbing plus heating system plus electrical system', and so on.

The requirements that any of these components should meet can also be evoked at appropriate times in the design process, and need not be specified in advance. Consideration of the heating system can evoke from the architect's long-term memory (or the appropriate reference handbooks) that the system should be designed to maintain a temperature of 70°, that the minimum outside temperature to be expected is −5°, that the heat transmission coefficient of the proposed sheathing is kBTU per hour per square foot per degree of temperature differential, and so on.

Design alternatives can also be evoked in component-by-component fashion. The subgoal of designing the heating system may lead the architect to consider various fuels and various distribution systems. Again, the source of these generators of alternatives is to be found in his long-term memory and reference facilities (including his access to specialists for helping design some of the component systems).

The whole design, then, begins to acquire structure by being decomposed into various problems of component design, and by evoking, as the design progresses, all kinds of requirements to be applied in testing the design of its components. During any given short period of time the architect will find himself working on a problem which, perhaps beginning in an ill-structured state, soon converts itself through evocation from memory into a well-structured problem. We can make here the same comment we made about playing a chess game: the problem is well-structured in the small, but ill-structured in the large.

Coordination of the design

Now some obvious difficulties can arise from solving problems in this manner. Interrelations among the various well-structured sub-problems are likely to be neglected or under-emphasized. Solutions to particular sub-problems are apt to be disturbed or undone at a later stage when new aspects are attended to, and the considerations leading to the original solutions forgotten or not noticed. In fact, such unwanted side effects accompany all design processes that are as complex as the architectural one we are considering. As a result, while the final product may satisfy all the requirements that are evoked when that final product is tested, it may violate some of the requirements that were imposed (and temporarily satisfied) at an earlier stage of the design. The architect may or may not be aware of the violation. Some other appropriate design criteria may simply remain dormant, never having been evoked during the design process.

The danger of inconsistencies and lacunae of these kinds is mitigated to some extent by that part of the architect's skill that is embedded in the over-all organization of his programme for design. Part of his professional training and subsequent learning is directed to organizing the process in such a way that the major interactions among components will be taken care of. Certain ways of dividing the whole task into parts will do less violence to those interactions than other ways of dividing it—a good procedure will divide the task into components that are as nearly 'self-contained' as possible (Alexander, 1964). Early stages of the design can also establish global parameters which then become constraints operating on each of the components to which they are relevant. Thus general decisions about 'style' can impose constraints on the stylistic decisions about particular portions of the house.

Much of the coordination of the various well-structured design sub-tasks is implicit—built into the organization of the whole process. To the extent that this is so, the local design activities are guaranteed to mesh into a reasonable overall structure. This means that the final product may be very much influenced by the order in which the design steps are taken up. As a result, differences in style between different designs can result as readily from the organization of the design process as from explicit decisions of the architect to specify one style or another. If the process calls for designing the façade before the floor plan, different kinds of designs will emerge than if the process calls for specifying the room arrangements before the façade (Simon, 1971).

The over-all design process

The design process sketched above can be composed from a combination of a GPS, which at any given moment finds itself working on some well-structured sub-problem, with a retrieval system, which continually modifies the problem space by evoking from long-term memory new constraints, new sub-goals, and new generators for design alternatives. We can also view this retrieval system as a recognition system that attends to features in the current problem space and in external memory (e.g. models and drawings), and, recognizing features as familar, evokes relevant information from memory which it adds to the problem space (or substitutes for other information currently in the problem space). The retrieval system, then, is capable of interrupting the ongoing processes of the problem solving system. A schematic flow diagram for a system with these characteristics is shown in Figure 1.

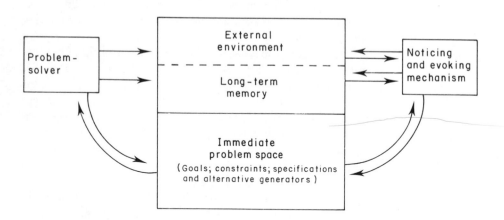

One might ask why the system is organized in this fashion—with alternation betwen problem-solving in a (locally) well-structured problem space and modification of the problem space through retrieval of new information from long-term memory. The answer revolves around the basically serial character of the problem-solving system. The problem-solving processes are capable, typically, of taking a few arguments as inputs and producing a small number of symbol structures as outputs. There is no way in which a large amount of information can be brought to bear upon these processes locally—that is, over a short period of processing. If a large

Figure 1
Schematic diagram of a system for ill-structured problems. It shows the alternation between a problem-solver working on a well-structured problem, and a recognition system continually modifying the problem space.

long-term memory is associated with a serial processor of this kind, then most of the contents of long-term memory will be irrelevant during any brief interval of processing.

Consider the alternative, of bringing all of the potentially relevant information in long-term memory together once and for all at the outset, to provide a well-structured problem space that does not change during the course of the problem-solving effort. One must ask 'bringing it together where?' Presumably it could all be assembled in some designated part of long-term memory. But to what purpose? Retrieval processes would still be necessary for the serial problem-solving processes to discover the inputs they needed at the time they needed them. To the outside observer the continuing shift in attention from one part of the assembled task information to another would still look like a series of transformations of the problem space.

In the organization described in Figure 1, there is no need for this initial definition of the problem space and task structure. All of the necessary definitory information is potentially available, but distributed through long-term memory. It is retrieved through two mechanisms: first, the normal sub-routine structure, which enables processes to call sub-processes and to pass input and output information from one to another; second, the evoking mechanism, which recognizes when certain information has become relevant, and proceeds to retrieve that information from long-term memory.

3.3. Design as an organizational process

If this description of how an ill-structured problem of design can be handled seems at all fanciful, its realism can be supported by comparing it with the description of complex design processes that take place in organizations. Let me quote at length the process, described by Sir Oswyn Murray some fifty years ago, of designing and producing a new battleship (Murray, 1923, pp. 216–17):

We start with the First Sea Lord and his Assistant Chief of Naval Staff laying down in general terms the features that they desire to see embodied in the new design—the speed, the radius of action, the offensive qualities, the armour protection. Thereupon the Directory of Naval Construction, acting under and in consultation with the Controller, formulates provisional schemes outlining the kind of ship desired together with forecasts of the size and cost involved by the different arrangements. To do this he and his officers must have a good general knowledge—in itself only attainable by close relations with those in charge of these matters—of the latest developments and ideas in regard to a great range of subjects—gunnery, torpedo, engineering, armour, fire-control, navigation, signalling, accommodation, and so on—in order to be reasonably sure that the provision included in his schemes is likely to satisfy the experts in all these subjects when the time for active cooperation arrives.

With these alternative schemes before them, the Sea Lords agree on the general lines of the new ship, which done, the actual preparation of the actual designs begins. The dimensions and shape of the ship are drawn out approximately by the naval constructors. Then the Engineer-in-Chief and his Departments are called in to agree upon the arrangement of the propelling machinery, the positions of shafts, propellors, bunkers, funnels, etc., and at the same time the cooperation of the Director of Naval Ordnance is required to settle the positions of the guns with their barbettes, and magazines and shell rooms and the means of supplying ammunition to the guns in action.

An understanding between these three main departments enables further progress to be made. The cooperation of the Director of Torpedoes and the Director of Electrical Engineering is now called for to settle the arrangements for torpedo armament, electric generating machinery, electric lighting, etc. So the design progresses and is elaborated from the lower portions upwards, and presently the Director of Naval Construction is able to consult the Director of Naval Equipment as to the proposed arrangements in regard to the sizes and towage of the motor boats, steamboats, rowing and sailing boats to be carried, as well as of the anchors and cables; the Director of the Signal Department as to the wireless telegraphy arrangements; the Director of Navigation as to the arrangements for navigating the ship, and so on. In this way the scheme goes on growing in a tentative manner, its progress always being dependent on the efficiency of different parts, until ultimately a more or less complete whole is arrived at in the shape of drawings and specifications provisionally embodying all the agreements. This really is the most difficult and interesting stage, for generally it becomes apparent at this point that requirements overlap, and that the best possible cannot be achieved in regard to numbers of points within the limit set to the contractors. These difficulties are cleared up by discussion at round-table conferences, where the compromise which will least impair the value of the ship are agreed upon, and the completed design is then finally submitted for the Board's approval. Some fourteen departments are concerned in the settlement of the final detailed arrangements.

The main particular in which this account differs from our description of the architectural process is in the more elaborate provision, in the ship design process, for coordinating the numerous criteria of design that are evoked during the process, and preventing criteria, once evoked, from being overlooked in the latter stages of the process. In the ship design process too the 'evoking' takes on an organizational form—it involves consulting the relevant specialists. The long-term memory here is literally a distributed memory, divided among the various groups of experts who are involved at one or another stage of the design process.

What is striking about Sir Oswyn's account is how well structured each part of the design process appears. We can visualize each group of experts, provided with the overall specifications, sitting down with their specific sub-problem of designing a particular system, and finding that sub-problem to be a relatively well structured task. Of course the more complex sub-problems may themselves be in numerous ways ill-structured until further subdivided into components (cf. the descriptions of the role of the Engineer-in-Chief and of the Director of Naval Ordnance).

Wherever complex designs are produced by organizations, and where, as a consequence, the design process must be partially externalized and formalized, we find descriptions of that process not unlike this description of ship design. An initial stage of laying down general (and tentative) specifications is followed by stages in which experts are called up ('evoked') to introduce new design criteria and component designs to satisfy them. At a later stage there is attention to inconsistencies of the component designs, and a search for modifications that will continue to meet most of the criteria, or decisions to sacrifice certain criteria in favour of others. Each small phase of the activity appears to be quite well structured, but the overall process meets none of the criteria we set down for WSPs.

3.4.   An intelligent robot

A different aspect of structure comes to the forefront when we consider the design of an intelligent robot capable of locomoting and solving problems in a real external environment. The robot's planning and problem-solving must be carried out in terms of some internal representation of the external environment. But this internal representation will be inexact for at least two reasons: first, it must abstract from much (or most) of the detail of the actual physical environment. (It surely cannot represent the individual molecules and their interactions, and it must almost always ignore details that are much grosser and more important than those at the molecular level.) Second, the internal representation includes a representation of the changes that will be produced in the external environment by various actions upon it. But for this prediction of the effects of operators to be exact would require an exact knowledge of the laws of nature that govern the effects of real actions upon a real environment.

The robot, therefore, will continually be confronted with new information from the environment: features of the environment which have become relevant to its behaviour but are omitted from, or distorted in, its internal representation of that environment; and changes in the environment as a consequence of its behaviour that are different from the changes that were predicted by the planning and problem-solving processes.

But this external information can be used by the robot in exactly the way that information evoked from long-term memory was used by the architect. The problem representation can be revised continually to take account of the information— of the real situation—so that the problem-solver is faced at each moment with a well-structured problem, but one that changes from moment to moment.*

If the continuing alteration of the problem representation is short-term and reversible, we generally call it 'adaptation' or 'feedback', if the alteration is more or less permanent (e.g. revising the 'laws of nature'), we refer to it as 'learning'. Thus the robot modifies its problem representation temporarily by attending in turn to selected features of the environment; it modifies it more permanently by changing its conceptions of the structure of the external environment or the laws that govern it.

3.5. Chess playing is an ISP

We can now return to the task environment of chess, and reinterpret our earlier comments—that in some respect chess appears to resemble an ISP rather than a WSP. To make the point still sharper, we consider a (over-simplified) chess program that has three principal components:

(1) a set of productions;
(2) an evaluator;
(3) an updating process.

The set of productions serves as a move generator. Each production consists of a condition part and an action part; the condition part tests for the presence of some configuration of pieces or other features in a chess position; the action part evokes a move that should be considered when the condition with which it is associated is present. Such a set of productions can also be viewed as an indexed memory, or discrimination and recognition net. When the presence of a condition is recognized by the net, the corresponding action is accessed in long-term memory. By the operation of this system of productions, each chess position evokes a set of moves that should be considered in that position.

The second component of the chess program is an evaluator that takes a move and a position as input and (recursively) makes a dynamic evaluation of that move. We will assume—without specifying exact mechanisms—that the dynamic evaluation converges. The evaluator produces a search tree that

---

*'Robots' that operate on synthesized worlds represented inside the computer, like the well-known system of T. Winograd, are not robots in the sense in which I am using the term here; for they do not face the issue that is critical to a robot when dealing with a real external environment—the issue of continually revising its internal representation of the problem situation to conform to the facts of the world. Thus the information given PLANNER, the problem solving component of Winograd's system, is a complete and accurate characterization of the toy world of blocks that the system manipulates. The accuracy of the information guarantees in turn that any theorem proved by PLANNER will be true of the block world.

constitutes its prediction of the consequences that might follow on any given move.

The third component of the chess program is the updating routine. After a move has been made and the opponent has replied, the updater brings the position on the board up to date by recording the moves, prunes the analysis tree, and turns control over again to the discrimination net.

Consider now a problem space consisting of the set of features recognizable by the productions, together with the moves associated with those features. Consider the subspace consisting of the features actually evoked by a given position, together with the moves associated with this smaller set of features. If we regard the latter, and smaller, space as the effective problem space for the program (since its processing during a limited period of time will be governed only by the productions actually evoked during that time), then the effective problem space will undergo continuing change throughout the course of the game, moving from one subspace to another of the large space defined by the entire contents of long-term memory. The problem faced by the chess program will appear just as ill-structured as the architect's problem or the robot's problem—and for exactly the same reasons.

3.6. Serial processors and ISPs

Our analysis has led us to see that any problem solving process will appear ill-structured if the problem-solver is a machine that has access to a very large long-term memory (an effectively infinite memory) of potentially relevant information, and/or access to a very large external memory provides information about the actual real-world consequences of problem-solving actions. 'Large' is defined relative to the amount of information that can direct or affect the behaviour of the processor over any short period of time; while 'potentially relevant' means that any small part of this information may be evoked at some time during the problem-solving process by recognition of some feature in the current problem state (the information available *directly* to the processor).

If we refer back to the original definition of WSP, we see that the present view of the boundary between WSP and ISPs derives from our insistence that notions of computability in principle be replaced by notions of practicable computability in the definition of WSP. But this shift in boundary has highly important consequences. It implies that, from the standpoint of the problem-solver, any problem with a large base of potentially relevant knowledge may appear to be an ill-structured problem; and that the problem-solver can be effective in dealing with it only if it has capabilities for dealing with ISPs. Conversely, it suggests that there may be nothing other than the size of the knowledge base to distinguish ISPs from WSPs, and that general problem-solving mechanisms that have shown themselves to be efficacious for handling large, albeit *apparently* well-structured domains should be extendable to ill-structured domains without any need for introducing qualitatively new components.

However well-structured the problem space in which a problem-solver operates, if it is to be capable of modifying that space as problem-solving progresses, it must possess means for assimilating the information it acquires from long-term memory, from problem instructions, and from the external environment. The next section discusses briefly the nature of the capabilities of these kinds that are required.

4.   ASSIMILATING NEW INFORMATION

When the problem space remains unchanged throughout the problem-solving process the assimilation of new information by the problem-solving system offers no particular difficulties. Any information that can be used belongs to one of the forms of information that are specified in the definition of the problem space. In fact, the only new information that can be acquired is information descriptive of new states that are reached in the course of the problem-solving search.

When the problem space is subject to modification during problem-solving, provision must be made for accepting and assimilating information from one or more of three sources: information evoked from long-term memory, information contained in problem instructions or additions and modifications to instructions, and information obtained through sensory channels from the external world.

4.1.   Information from long-term memory

Information evoked by the production–recognition system from long-term memory should not create particular difficulties of assimilation. The forms in which such information is stored, the processes for retrieving it and incorporating it in the redefined problem space are all part of the problem-solving system, broadly construed.

This does not mean that we have had much actual experience in constructing and using such semantic information stores. Perhaps, when we come to have such experience, difficulties will emerge that cannot now be anticipated. Nevertheless, the designer of the problem-solving system controls the format in which information is to be stored in long-term memory. The storage scheme can put that information into a relatively simple, general, and homogeneous format (e.g. a network of description list structures—commonly referred to as a coloured directed graph); and the system can be provided with relatively simple, general, and homogeneous processes for searching for information and retrieving it from the store. Interfacing with an environment whose design is under our control (memory search) is always several orders of magnitude easier than interfacing with a given external environment (perception).

4.2. Information from instructions

Tasks presented to a problem-solver through natural language instructions pose the difficult initial problem of understanding the instructions, that is, of generating from them a well-structured (or ill-structured) statement of the problem. A general discussion of understanding natural language would take us far afield from our main concerns here, and must be omitted from the present paper. The reader is referred to Simon (1972), Siklóssy and Simon (1972), chapters by Bobrow and Raphael Minsky (1969) and Winograd (1973).

Hayes and Simon (1974) have constructed a system that reads problem instructions in a natural language, and constructs from them a problem representation in the form of input appropriate for a problem-solving system like GPS. Their program carries out the major steps in translating relatively simple ISPs into WSPs.

4.3. Information about the external world

An ability to assimilate information about the external world—either information about the effects of the problem-solver's actions, or information about autonomous changes in that world, or both—is the earmark of those problem-solvers we have called 'robots'. Here we are concerned with the problem-solver's perceptual or pattern-recognizing capabilities—another topic that falls outside the scope of the present paper. The outer limits on acquisition of such information are defined by the primitive sensory discriminations of which the problem-solver is capable, but in the short run the higher-level concepts already developed and stored may pose the most severe problems of assimilating new information.

In the cases both of information from instructions and information about the external world, the acquisition process involves a continual interaction between the incoming raw data and programs and data already stored in the problem-solver. This interaction is both a vital aid to, and a limitation upon, the understanding process. It is an aid because it fits the new information to formats and structures that are already available, and adapts it to processes for manipulating those structures. It is a limitation, because it tends to mould all new information to the paradigms that are already available. The problem-solver never perceives the *Ding an sich*, but only the external stimulus filtered through its own preconceptions. Hence the acquisition process exercises a strong influence in the direction of conserving existing problem formulations. The world as perceived is better structured than the raw world outside.

5. IMPLICATIONS FOR ARTIFICIAL INTELLIGENCE

Molière's hero discovered that he had been speaking prose all his life without knowing it. Our analysis here implies that throughout the history of artificial intelligence, computer problem-solving programs have, also unknowingly, been handling many aspects of problem-solving that are usually regarded as ill-structured. Several examples of programs now fifteen years old can be cited.

The early NSS chess program (Newell *et al.* 1958) contained a number of independent move generators, each associated with a particular sub-goal (development, centre control, King safety, etc.). A move generator was activated by the presence in the chess position of features relevant to the goal in question. When evoked, the generator proposed one or more moves for advancing the goal in question. These moves were evaluated by dynamic analysis which again was sensitive to features noticed in new positions as they arose during the analysis. Hence the over-all organization of the program was quite close to that of the hypothetical program we described earlier.

The proposals for using a planning method in the General Problem Solver can also be interpreted as a way of handling problems that are not completely well-structured. Planning was done by abstracting from the detail of a problem space, and carrying out preliminary problem solving in the abstracted (and consequently simpler) space. But the plan then had to be tested by trying to carry it out in the original problem space. The detail of the plan was thereby elaborated and its feasibility tested. The relation between the abstract planning space and the original problem space was quite analogous to the relation we have discussed between an internal problem space of a robot and the external real world in which the robot performs.

What has been absent from (or at least not prominent in) schemes like these is a very large long-term memory of potentially evocable information that can be used to bring about repeated modifications in the problem space. The recent growth of interest in semantics, in the design and construction of semantic nets, in the interpretation of instructions, and in interfacing robots with the external world are all movements in the direction of enriching our arsenal of artificial intelligence methods along the dimensions that will become increasingly important as we move toward more comprehensive schemes for handling ill-structured problem-solving. Our analysis gives us reason to be optimistic that progress will not require us to introduce mechanisms that are qualitatively different from the ones already introduced in artificial intelligence schemes—mechanisms with which we have already had some limited experience.

The boundary between well-structured and ill-structured problem-solving is indeed a vague and fluid boundary. There appears to be no reason to suppose that concepts as yet uninvented and unknown stand between us and the fuller exploration of those problem domains that are most obviously and visibly ill-structured.

Acknowledgments

This research was supported in part by Research Grant MH-07722 from the National Institute of Mental Health and in part by the Advanced Research Projects Agency of the Office of the Secretary of Defense (F44620-70-C-0107) which is monitored by the Air Force Office of Scientific Research. I am grateful to Dr Aaron Sloman and Prof. Saul Amarel for helpful comments on an earlier draft of this paper.

REFERENCES

Alexander, C. (1964), *Notes on the Synthesis of Form*, Harvard University Press, Cambridge, Mass.

Hayes, J. R. and Simon, H. A. (1974), 'Understanding written problem instructions', *Proc. 9th Carnegie Symp. on Cognition*, L. W. Gregg (ed.).

Kling, R. D. (1971), 'A paradigm for reasoning by analogy', *Proc. 2nd Intern. Joint Conf. on Artificial Intelligence*, British Computer Society, London, pp. 568–85.

Minsky, M. (ed.) (1969), *Semantic Information Processing*, MIT Press, Cambridge, Mass.

Murray, Sir Oswyn A. R. (1923), 'The administration of a fighting service', *J. Public Admin.* **1** (July), 216–17.

Newell, A. (1969), 'Heuristic programming: ill structured problems', *Progress in Operations Research*, Vol. 3, J. Aronofsky (ed.), Wiley, New York, pp. 360–414.

Newell, A. and Simon, H. A. (1972), *Human Problem Solving*, Prentice-Hall, Englewood Cliffs, N.J.

Newell, A., Shaw, J. C. and Simon, H. A. (1958), 'Chess-playing programs and the problem of complexity', *IBM J. Res. Develop.*, **2**, 320–335.

Pitrat, J. (1966), 'Réalisation des programmes de démonstration des théorèmes utilisant des méthodes heuristiques'. Doctoral thesis, Faculty of Science, University of Paris.

Reitman, W. R. (1964), 'Heuristic decision procedures, open constraints, and the structure of ill-defined problems', *Human Judgments and Optimality*, M. W. Shelley and G. L. Bryan (eds), Wiley, New York, pp. 282–315.

Reitman, W. R. (1965), *Cognition and Thought*, Wiley, New York.

Siklóssy, L. and Simon, H. A. (1972), 'Some semantic methods for language processing', *Representation and Meaning*, H. A. Simon and L. Siklóssy (eds), Prentice-Hall, Englewood Cliffs, N.J.

Simon, H.A. (1971), 'Style in design', *Proc. 2nd Ann. Environ. Design Res. Assoc. Conf.*, 1–10, October 1970, J. Archea and C. Eastman (eds), Carnegie-Mellon University, Pittsburgh, Pa.

Simon, H.A. (1972), 'The heuristic compiler', *Representation and Meaning*, H.A. Simon and L. Siklóssy (eds), Prentice-Hall, Englewood Cliffs, N.J.

Winograd, T. (1973), 'Understanding natural language', *Cognitive Psychol.*, **3** (1), 1–191.

Part Three

The Nature of Design Activity

Introduction

The third main area of study and research in design methodology has been the investigation of what it is that designers actually do when they are designing. In general, the intention has been to try to develop an objective understanding of how designers design, which might then in turn lead to the development of improved design procedures. This line of research treats design activity as a 'natural phenomenon' of human behaviour, and therefore relies heavily on methods of enquiry drawn from psychology and the study of human cognitive performance. As with all such methods, there are obvious limitations to studying what goes on inside someone's head, and the researchers have to construct inferences from the behaviour and the reported thoughts and mental operations of their subjects.

The simplest way of studying designers' behaviour is to ask them to recall what they did when they were designing. Whilst this lacks the controlled experimental precision of laboratory observations, it does produce very 'rich' data which carry the intrinsic interest of reflections on real-world designing. Darke, in her paper on 'The primary generator and the design process', goes further and suggests that it is undesirable to confine the analysis of human behaviour to strictly 'scientific' methods. Her own research is based on interviews with architects who had designed various housing schemes. Of course there are severe problems with this approach, too, and she notes that, for example, her respondents might have faulty memories, might post-rationalize on their activities, and almost certainly would have difficulties in describing non-verbal design processes in words.

Darke's starting point was a dissatisfaction with the early systematic design procedures which presumed an objective analysis–synthesis approach to designing. She uses the evidence from her interviews to support an alternative approach of 'conjecture–analysis' (derived from Hillier *et al.*, see Chapter 4.1), and to suggest that, in addition, designers rely firstly on the formulation of a 'primary generator'. Very early in the design process the designer imposes (or identifies) a particular generating concept or limited set of objectives. 'These objectives form a starting point for the architect, a *way in* to the problem; he does *not* start by listing all the constraints.' This primary generator helps the designer (to make the 'creative leap'?) across the 'rationality gap' between the problem information and a solution concept.

The use of a primary generator—often some aspect of the site which was to be built on—was characteristic of all the architects Darke interviewed. She concluded that this is a necessary feature of the design process, because designers 'have to find a way of reducing the variety of potential solutions to the as yet imperfectly understood problem, to a small class of solutions that is cognitively manageable'. The solution class is further narrowed by proposing one particular solution concept (the 'conjecture') which is then tested against the requirements and constraints of the problem, thus contributing to a fuller understanding, or analysis, of the problem. Darke therefore concludes that 'the analysis–synthesis model would seem to be refuted as a method which can readily be used in practice'.

An attempt to develop a more objective method of studying problem-solving behaviour—protocol analysis—has been mentioned by Simon in Chapter 2.4, and he has been one of the leading authorities to use and develop this method. It was the method used to study design behaviour by Akin in his 'An exploration of the design process'. His purpose was to study 'intuitive-design'—i.e. the natural behaviour of human designers—which he distinguishes from the more systematic procedures of 'design-methods' and the computer aided techniques of 'machine-design'. He suggests that a better understanding of intuitive-design will not only enable appropriate design-methods and machine-design procedures to be formulated, but also could inform normal design practice and improve design education.

The main goal of Akin's research was to 'disaggregate' the design process, i.e. to break it down into its component parts, and he chooses protocol analysis as the most appropriate technique for this. Protocol studies rely on the subject reporting aloud what he or she is thinking and doing whilst tackling the given experimental task. In contrast with more controlled experiments in human behaviour, Akin suggests that protocol studies are particularly suited to the study of behaviour which is poorly understood, and can very usefully be applied in the case of the ill-defined nature of design problems. However, for his own study he modifies the established techniques of protocol analysis and adopts the use of 'Plans' as descriptions of design operations. 'Plans are hierarchical processes that control the order in which a sequence of operations is to be performed.'

Akin's empirical study of the design process involved setting up quasi-laboratory conditions for recording the behaviour and the spoken thoughts of a designer performing a design task. The particular instance reported is that of an architect given a moderately sized design problem, i.e. the design of a single-person dwelling. The study therefore uses a real designer and a realistic design problem, but in controlled conditions for the sake of reliable observation and recording, and with the imposed artificial condition of the designer being asked to report his thoughts aloud.

On completion of the task (to sketch-plan stage in this case) the designer's behaviour is analysed by the experimenter and transformed into formalized statements of recurring behaviour patterns—i.e. into 'Plans' or 'Schemata' in the terms used by Akin, who identifies eight cognitive schemata used by the designer: instantiation, generalization, enquiry, inference, representation, goal-definition, specification, and integration.

From his analysis Akin identifies a hierarchy of design strategies, beginning with 'setting-up' the design context and then searching for sub-solutions. He likens the designer's general solution-search strategy to 'hill-climbing', i.e. the designer starts more or less at any point in the solution space and then tries to move from there towards a local optimum by pursuing a sequence of small improvements.

Akin concludes that some long-held views of the design process do not reflect normal design behaviour. Like Darke, he suggests that the systematic procedure of analysis–synthesis–evaluation seems inappropriate: 'One of the unique aspects of design behaviour is the constant generation of new task goals and redefinition of task constraints. Hence "analysis" is a part of virtually all phases of design. Similarly "synthesis", or solution development, occurs as early as in the first page of the protocol.' He suggests that normal design behaviour is to start with a broad, top-down approach to the task, and that designers realistically attempt to 'satisfice' rather than to optimize solutions. Akin's view is that the development of design methods must be based on a recognition of designers' normal ways of working if those methods are to integrate successfully.

A more controlled experimental approach to studying designers' problem-solving behaviour was taken by Lawson in his investigation of 'Cognitive strategies in architectural design'. Lawson starts with a generalized view of architectural design, in which 'The architect's primary and central task is to produce a "concrete" three-dimensional structure of space and form to accommodate an abstract structure of related human activities.' For his experimental purposes he required a controllable, 'model' design activity, and so he devised a task which required subjects to produce a spatial configuration of elements so as to try to achieve some given goal and to satisfy some (initially unknown) rule. The task was based on selecting and arranging coloured blocks of different shapes so as to try to maximize the amount of one certain colour showing around the outside faces, with an undisclosed rule requiring certain blocks to be present. The subjects could enquire whether any proposed solution was acceptable, i.e. satisfied the rule, and in this way could attempt to discover the rule. However, they were required only to produce their best acceptable solution, without necessarily explicating the rule.

In his main experiments Lawson compared the performances of fifth-year architectural students and fifth-year science students on the various tasks which can be devised with the coloured blocks. Each student's performance was measured in terms of the amount of the required colour showing on the external faces. The mean performance scores of both student groups were very similar, but their performances were different for different types of problem rules: the architects did better with 'conjunctive' rather than 'disjunctive' rules, whilst the scientists were the reverse. (A 'conjunctive' rule required that two specific blocks must both be included; a 'disjunctive' rule required that either or both of two blocks must be included.) Lawson identifies two types of error which would prevent a subject achieving the maximum possible score: planning errors, in which there was simply a non-optimal arrangement of the blocks, and structural errors, in which there was (usually) an assumption of a more constraining rule than actually did apply. As might be expected, the architects made fewer planning errors, but made more structural errors than the scientists.

The more interesting aspect of Lawson's results is in his analysis of the differences in problem-solving strategies between the two groups. He discovered that, in general, the scientists were selecting blocks in procedures which were aimed at uncovering the problem structure (i.e. the hidden rule), whereas the architects' procedures were aimed at generating a sequence of high-scoring solution attempts until one proved acceptable. Lawson calls these two different problem-solving strategies 'problem-focused' (scientists) and 'solution-focused' (architects). The implication is that designers' methods are quite different from scientists' methods. The architects had learned that the most successful (or practicable) way to tackle design problems is by proposing solutions to them; in this way one discovers more about the problem and what is an 'acceptable' solution to it. The scientists had learned to analyse a problem and to attempt to discover its 'rules' before proposing a solution to it. In other words, designers problem-solve by methods of synthesis, whereas scientists problem-solve by methods of analysis.

Lawson suggests that 'If the conclusions drawn from this empirical work are correct then this raises some difficult questions to be answered by the proponents of the view that design methodology can be understood as a special subset of scientific methodology.' The two approaches must be difficult to reconcile if they start from such different orientations. Lawson argues that his conclusions lend support to the 'wicked problems' view—i.e. that design problems can never be completely described and therefore never susceptible to a complete analysis. In such a case, therefore, 'it seems quite reasonable to suppose that designers would evolve a methodology which does not depend on the completion of problem analysis before synthesis can begin'. That is, designers' habitual methods are quite appropriate to their habitual problems.

Several different experimental and observational methods of studying designer behaviour are reported by Thomas and Carroll in their paper, 'The psychological study of design'. They used a variety of methods to study designing, ranging from reflections on their own experiences to observations of others—and the latter observations ranged from recording real design dialogues between clients and designers to setting structured design problems to groups of selected subjects. Their interpretation of 'design' is also very wide—from computer software design to letter-writing.

From their observations of 'design dialogues', Thomas and Carroll disaggregate the design process into six states or operations: goal statement, goal elaboration, solution outline, solution elaboration, solution testing, and agreement or rejection of solution. They view the overall process as a 'dialectic interaction between the information in the client's head and that in the designer's head'. What distinguishes the designer's contribution is his knowledge about the relationships between partial goals and partial solutions. However they also found, from software design studies, radically different designs being proposed by different designers who were making different goal assumptions.

In some of their more controlled experiments Thomas and Carroll found that subjects using a design aid which was meant to help explicit problem formulation actually produced worse results than those with no aid. But the experiment did show again that people often misunderstand problem goals. In an experiment which required subjects to design an 'improved' chair, Thomas and Carroll found that a design aid which was simply a random word list did seem to help produce more creative solutions. They also set up experiments in which subjects were given design problems which were isomorphic in structure, although one was a spatial problem (office planning) and one was a temporal problem (production scheduling). They found that subjects solved the spatial problem better and quicker than the temporal problem, except when given a simple representation aid (a matrix), in which case there were no differences.

Thomas and Carroll conclude from their wide range of studies that 'Our original suspicion that there were similarities in design regardless of the domain has been strengthened.... Activities as diverse as software design, architectural design, naming and letter-writing appear to have *much* in common.' They also make some interesting observations on how they modified their original assumption that designing is a form of problem-solving for ill defined problems to the opinion that designing is essentially a *way of looking at* a problem. It is therefore not exclusive to a particular *type* of problem. As an example, theorem-proving can be *viewed* as designing if, say, the requirement to stay inside certain formal rules is relaxed and if creativity is allowed; then, although the initial goal may have seemed to be to prove a theorem, actually it was 'to find out something interesting'. Conversely, they argue, house design could be treated as a problem to be solved by applying a set of standard rules to the stated requirements; but in this case what is happening is *not* design. Therefore whether a task is a design task depends on the problem-solver: certain problems can be treated either as ill-defined or well-defined.

It is interesting that Thomas and Carroll conclude that designing is a generalizable form of problem-solving which can be applied in a wide variety of contexts, since this was one of the original assumptions underlying design methodology. If Thomas and Carroll are correct, then the general findings from the other authors' studies of architectural designing should be equally relevant to other fields, such as engineering design or industrial design. A consistent view of how architects design does emerge from the three other authors, even though each takes a different approach to studying designer's behaviour. For example, it seems clear that architects have a 'solution-focused' approach to design and that they begin to generate solution concepts very early in the design process. This is presumably typical of ill-defined problem-solving approaches; because the problem is ill-defined there is an inevitable emphasis on the early generation of a solution so that an understanding of the problem-and-solution can be developed. An ill-defined problem is never going to be completely understood without relating it to a potential solution.

The other general conclusion which is stressed by most of the authors here is that most systematic procedures are ill-matched to the conventional design process. Systematic procedures tend to assume or require an extensive phase of problem analysis, which seems an unrealistic approach to ill-defined problems. Darke, Akin, and Lawson all criticize the systematic analysis–synthesis procedure, in the light of their observations of how designers design. However, it would be tautologous simply to argue that conventional designing is unlike systematic designing; the systematic procedures were developed specifically to be a change from conventional design practices, which were seen to be inadequate for the complexity of the tasks facing modern designers.

Architects, in particular, have been criticized for the quality of their design solutions, and so it does seem as though there is a need to improve on their design procedures. Criticism of systematic procedures, therefore, should not be taken as arguments justifying the conventional behaviour of designers.

FURTHER READING

Two early observational studies of designer behaviour were published in Moore, G. T. (ed.) (1970), *Emerging Methods in Environmental Design and Planning*, MIT Press, Cambridge, Mass. One was a report of a typical building design project: Krauss, R. I., and Myer, J. R., 'Design: a case history'; and the other was a protocol analysis of subjects attempting constrained design problems: Eastman, C. M., 'On the analysis of intuitive design processes'.

Lawson's work is developed more fully in Lawson, B. R. (1980), *How Designers Think*, Architectural Press, London.

A review paper is provided by Yeomans, D. (1982), 'Monitoring design processes', in Evans, B., Powell, J., and Talbot, R. (eds), *Changing Design*, Wiley, Chichester.

3.1 The Primary Generator and the Design Process

Jane Darke

RESEARCH DESIGN AND METHODOLOGY

Often in research a topic which was expected to be of minor importance when the research was conceived acquires a greater significance in the course of the project. The project which gave rise to the 'findings' reported here was not originally envisaged as 'design methods research'. It was intended to evaluate the design of some recent local authority housing schemes, with two main areas of interest. Firstly, it was intended to test current assumptions on the undesirability of high rise and high density, with the hypothesis that a well thought-out scheme at high rise and/or density could be satisfactory as a scheme at lower rise or density. (This hypothesis, it is worth stating, has not been supported by the results since high densities do seem to cause problems, although there is not a simple or automatic relationship between higher densities or high buildings and decreased satisfaction.)

Secondly, it was intended to see whether during the design of housing the architect has in mind an image or expectations regarding the future user, and if so whether this corresponds with the reactions of the actual user, his or her concerns and requirements regarding the home environment. The implicit hypothesis, that an environment seen as unsatisfactory by the user might result from an inaccurate perception of the user by the architect, has been difficult to test because of the very generalized nature of the architects' images of their users.

Originally published in *Design Studies*, **1** (1) (1979), 36–44. Reproduced by permission of Butterworth and Company (Publishers) Ltd.

The research procedure first required the selection of estates to study. These were chosen to minimize variation in factors extraneous to the research design, and at the same time to exemplify contrasts in their architectural design, especially in rise and density. It is not necessary to recount here the process of selection in detail; the outcome was that five estates were chosen (Table 1), all in London.

At least one of the architects of each scheme was interviewed at length and the interview taped. They were asked first to talk about their views on the design of housing in general and how these views had changed over time, then about the job history of the chosen scheme. In particular they described the evolution of the design, the existence or otherwise of an image of the future user, and the source(s) of any image of this kind. They were asked whether the design was the work of a single designer or a team, and if by a team, how they worked together. There were questions on disagreements with clients or superiors, and on any changes the designer(s) would have made if faced with the same design problem again. The question wording and order were not standardized and the intention was for a conversational rather than an interrogational atmosphere. It was found that the designers were generally very willing to talk about their schemes and appeared to enjoy doing so, as if talking about a favourite child. These interviews were carried out in 1975; the schemes had been designed a few years earlier.

The other part of the fieldwork consisted of interviews with a sample of the residents on each estate. The details of these

Table 1

Estate	Location	Client (London Borough)	Architects	Interview with	No. of dwell-ings	Density (persons per acre)	Height	Ref.
Dawson's Heights	Dulwich	Southwark	Borough Architect	Kate Mac-intosh	296	90	up to 12 storeys	*Archit. J.* 25/4/73
Linden Grove	Camber-well	Southwark	Neylan & Ungless	Michael Neylan	41	96	up to 3 storeys	*Archit. J.* 9/11/71
Kedleston Walk	Bethnal Green	Greater London Council	Douglas Stephen & Partners	Douglas Stephen	58	136	2 and 4 storeys	*Archit. J.* 1/11/72
Marquess Road	Islington	Islington	Darbourne & Darke	John Dar-bourne & Jeremy Lever	991	200	mainly up to 5 storeys	*Archit. Rev* Sept. 74
Pollards Hill	Mitcham	Merton	Borough Architect	Richard MacCormac	850	116	all 3 storeys	*Archit. Rev* April 71 *Archit. Des.* Oct. 71

interviews and their analysis need not concern us here. It is necessary, however, to consider the problems of methodology that arise in the analysis of the architect interviews. These bring the researcher face to face with the endemic problem of 'social sciences': how far can sociology use scientific methods to understand and interpret human behaviour? Is it permissible to use shared understandings to interpret subjects' comments? Or should the researcher try to stand outside their situation and avoid making assumptions about subjective meanings?

In practice the latter course is impossible; the researcher has to use his or her subjective judgment and to realize that an analysis that was confined to 'scientific' methods would miss the most interesting points. The architects were interviewed by a researcher who teaches in a school of architecture; they assume she understands current issues in housing design: the swing against high rise, the constraints of the cost yardstick, schemes that have been influential, etc. They make passing references, assuming the interviewer shares their frame of discourse sufficiently to understand them. Scientific detachment is not a helpful stance here, except in so far as the researcher must, in understanding the respondent's assumptions, also stand outside them and understand his reasons for making them.

Thus 'social' methods offer a gain over 'scientific' methods of analysis. To quote Aron (1964): 'Human behaviour presents an intrinsic intelligibility which depends on the fact that men are endowed with consciousness, with thought.' This willing embrace of subjective methods distinguishes this research from much of the large volume of past and ongoing design methods research.

THEORIES ON DESIGN METHOD

In the early 1960s a fruitful new approach to design methodology was generated by the realization that 'design' as a process was common to various fields: the several specialisms within engineering, industrial design, architecture, planning, and so on. Insights were gained from the study of approaches to solving simple problems in laboratory conditions where scientific research methods could be applied. (Eastman, 1970; Lawson, 1972.) Others attempted to 'design' a design process from first principles, a process referred to by Hillier *et al.* (1972) as 'metadesign'. A failing of the unified approach was that it paid little attention to the actual process of design as it occurred in 'real' situations. Some rather unfruitful attempts were made to observe designers at work but it seems to the present author that the research material necessary to understand the design process is not a set of sketches but a knowledge of the mental process the designer goes through. Observation of sketched and written output is a curious way of obtaining such material. Asking designers to recall their own processes would seem *prima facie* to get closer to the truth about such processes, albeit in a less verifiable form.

This method, of course, has problems of its own. Some of the architects interviewed in the present research found it difficult to describe a non-verbal process in words. Other problems include faulty recall, and post-rationalization by architects describing the process after the event. In this research it has been found necessary to treat the architects' accounts as *if* they were accurate summaries, bearing in mind that over-simplification and so on may have occurred. Such over-simplification, if it has occurred, may give rise to a simplified model of the design process, but such simplification is an aid to clarity at this stage, and can always be elaborated if and when more sophisticated research techniques yield the necessary evidence.

Some researchers will also dislike the fact that studying design methods in use reintroduces a differentiation in the act of designing between those in different fields, although of course as research evidence becomes available on these fields this could lead to an improved new meta-theory.

The view of the design process that informed most of the research of the sixties was based on an analysis-synthesis model. This simple dichotomization had many variants, involving elaboration of the main stages and often involving feedback loops; see for example Archer (1969) and Maver (1970). In many cases these models were derived from the design processes of designers in other fields, e.g. engineering or industrial design. The following description is fairly typical of the methods being taught at the more analytically minded schools of architecture in the sixties, although it was often expressed in diagrammatic form rather than in prose.

It was recognized that there were many factors to be considered in any design problem, some quantifiable and others 'subjective', although the non-quantifiable factors were progressively being transmuted, through research, into quantifiable form. One hoped-for consequence of this would be the possibility of transferring much of the process to the computer, which would not be limited by preconceptions and would thus produce a better solution. The designer was to start by exhaustively listing the relevant factors, then to consider the interactions between these factors and to set performance limits on those factors that could be so treated. Only then was he to start synthesizing requirements to generate a form, starting with clusters of related factors. It was to be hoped that the synthesis of various factors would almost automatically generate a form, with minimal need for the designer to exercise subjective judgments; subjectivity was felt to be full of risks, a threat to a good solution (see for example Whitehead and Eldars, 1964).

The fact that practising designers, by and large, did not use such a method was not seen as a drawback, but merely as a sign of their backwardness. Existing solutions were seen as 'bromide images' (Chermayeff and Alexander, 1963) that hindered the development of better solutions. Examples where the *use* of such a method has been described are few, but see Alexander (1963) and Hanson (1969). It is significant that Alexander has now rejected this approach and that Hanson admitted that the method was cumbersome and that he did not intend to use it again.

As research and thinking on design methods proceeded there was more recognition of the complexities of the process. Lawson (1972) and Broadbent (1973), among others, have outlined the development of the field. Wehrli (1968) identified a range of problem-types, from the puzzle, with a single solution, to the doubly open-ended problem, where the number of potential solutions is infinite and multiple solutions are sought. Different methods are appropriate at different levels of complexity. Individuals might differ in their approach to design. Lawson (1972) identified two contrasting styles of operation, the problem-focused and the solution-focused. In solving an experimental, design-like problem, science students more often adopted a problem-focus, which involved learning as much as possible about the structure of the problem before attempting a solution. The use of a solution-focus was more characteristic of students of architecture. They learnt about the problem by trying a solution and seeing where it went wrong. Both groups performed the task equally well, although the types of error were different for the two groups.

The same author discriminated between the various constraints influencing a design. They were cross-classified as being either internal or external to the design of the building itself, being imposed by designer, client or third party, and being implicit or explicit. These distinctions will be applied to the case material later in the paper.

A NEW MODEL

These refinements certainly represented a conceptual advance on previous thinking, which had concentrated on what was thought to be common to all design. However, there was growing doubt as to whether the analysis-synthesis model and its elaborations could still provide a satisfactory framework into which the new thinking could fit. With hindsight, this can be seen in Kuhnian terms as indicating an increasing need for a new paradigm (Kuhn, 1962). Hillier *et al.* (1972) supply such a paradigm. After clearly showing the inadequacy both of the image of the design process and the perceived role of research in much of the thinking of the sixties, they propose the replacement of the analysis-synthesis model with one of *conjecture–*

analysis. To quote excerpts from their paper which are relevant to the present paper:

only by prestructuring any problem, either explicitly or implicitly, can we make it tractable, to rational analysis or empirical investigation ... design is *essentially* a matter of prestructuring problems either by a knowledge of solution types or by a knowledge of the latencies of the instrumental set in relation to solution types, ... this is why the process of design is resistent to the inductive-empiricist rationality so common in the field. A complete account of the designer's operations during design would still not tell us where the solution came from.

Hence, design is seen as a process of 'variety reduction' with the very large number of potential solutions reduced by external constraints and by the designer's own cognitive structures,

used by the problem solver in order to structure the problem in terms in which he can solve it. ... There is also a very practical reason why conjectures of approximate solutions should come early on. This is that a vast variety of design decisions cannot be taken—particularly those which involve other contributors—before the solution in principle is known ... conjecture and problem specification thus proceed side-by-side rather than in sequence.

The present author's analysis of her interviews with architects supports this proposed model with evidence. Preliminary analysis of these interviews was initially towards the inductive end of the inductive-deductive spectrum, in that the researcher avoided making assumptions about whether the subjects could be expected to have a common design process, and about the nature of any such process. Yet the conjecture–analysis model is itself applicable at a meta-level to the process of 'analysing' (the language is deficient here) the interviews: the concept of the primary generator can be seen as a *conjecture*, which was then tested against the case material. This material, as will shortly be described, supported the conjecture. The idea of a primary generator was found to be a useful way of conceptualizing a particular stage in the design process; that stage that precedes a conjecture. Therefore an elaboration of Hillier's model is proposed, to one of:

generator–conjecture–analysis

The interviews with architects showed that the use of a few simple objectives to reach an initial concept was characteristic of these architects' approaches in design. The greatest variety reduction or narrowing down of the range of solutions occurs early on in the process, with a conjecture or conceptualization of a possible solution. Further understanding of the problem is gained by testing this conjectured solution.

Clearly in some cases where architects have described their own process of design (well-known examples include Spence on Coventry Cathedral, Utzon on the Sydney Opera House, and Lasdun on the National Theatre) a visual image came very early in the process. In other cases it appears that a certain amount of preliminary analysis takes place before the visual concept arises. It seems normal, however, for there to be a 'rationality gap': either the visual concept springs to mind *before* the rational justifications for such a form, or the analysis does not dictate *this particular* concept rather than others.

The concept or objective that generates a solution is here called the '*primary generator*'. It can in fact be a group of related concepts rather than a single idea. These objectives form a starting point for the architect, a *way in* to the problem; he does *not* start by listing all the constraints. Any particular primary generator may be *capable* of justification on rational grounds, but at the point when it enters the design process it is usually more of an article of faith on the part of the architect, a designer-imposed constraint, not necessarily explicit. Among the architects interviewed various objectives served as primary generators: to express the site, to provide for a particular relationship between dwelling and surroundings, to maintain social patterns, and the like. These will be looked at in greater detail later on.

The broad requirements of the client, at this stage in a housing scheme typically just a target density and mix of dwelling sizes, are used along with the designer-imposed primary generator in arriving at an initial conjecture or concept. The designer has been aware all along that there are several detailed requirements to be met by the design, but performances on these parameters are not specified in advance. Once the initial concept has been generated it is tested against these various requirements and modified if necessary; the performance levels with respect to particular requirements are decided interactively, in the light of the effect on the emerging concept and on other parameters. Naturally this process is often spiral or iterative in character; for example in housing design there is frequent switching between considerations of dwelling type plans and considerations of site layout, as each of these has important implications for the other. The conjecture is not rejected unless there is a fairly glaring mismatch between it and the detailed requirements. Probably the main difference between the practising architect and the student is that the former has the experience of solution types required for a realistic conjecture. A frequent problem in a school of architecture is the student who has a limited stock of generating ideas which he attempts to apply to every problem without considering whether they are appropriate.

The concept of the primary generator will become clearer with examples of its operation in the design of the schemes in the sample. We should first clarify its relationship with the first conceptualized image, the 'conjecture' in the terms of Hillier *et al.* The term 'primary generator' does not refer to that image but to the ideas that generated it: in the case of Coventry cathedral, say, the idea that the altar must be a focus, allied to themes of a phoenix arising from the ashes and so on. The primary generator will be a component of the designer's 'cognitive structures' referred to earlier. By becoming aware of ideas that are acting as generators, the designer may be able to evaluate them and widen their range if necessary.

EVIDENCE FROM THE CASE STUDIES

The job architect for Dawson's Heights was Kate Macintosh, then at the London Borough of Southwark. Her scheme was in fact the winner of a small intra-office competition for the site. The site is a hilltop in a suburban area, and the design is of two tall blocks with deck access, with a romantically stepping-down profile. The highest points of the two buildings are offset, and there are setbacks on plan so that the effect is that the ends of the two blocks curve inwards to give a sense of enclosure.

The designer had two major preoccupations at the start of the process: to express this unique site and to allow for difficult soil conditions. To quote from the transcript:

> I: 'Would you like to talk about the job history—the way you arrived at the design, this sort of thing'.
>
> KM: 'Well, obviously the site is a very unusual one, in London, and I always have been one of the romantic school that think you should try to express the unique quality of the site, if it has a unique quality, and that site undoubtedly has, so that was the main starting off point. And of course the other peculiar fact, apart from the fact that ... you get magnificent views in both directions ... was the fact that the hill is unstable, and only the top third, roughly speaking, was buildable economically, and we were advised by the London University soil experts that even if you put a single storey garage down you'd have to pile'.

Note that the site is spontaneously described as 'the main starting off point'. The second factor, the soil strength, was not a designer-imposed constraint but was obviously a major determinant of the solution. A wish to respect the scale of the neighbourhood prompted the use of a stepping-down profile; and the objective of making best use of sun and views suggested the staggering of the highest points in each block, 'as each "Leviathan" looks across the tail of its brother'.

Once the concept of stepping-down tall blocks had been established there was a great deal to be worked out in detail: whether to use a split-level plan, the dwelling types and how they interlocked, etc. This process clearly fits the 'generator–conjecture–analysis' model. Another factor should be mentioned, perhaps as a negative generating factor: the *lack* of a presumption *against* flats, by a designer brought up in a Scottish city rather than in England, and who had worked in Scandinavia where flats are common.

Site was again the starting point for Michael Neyland, a partner in the small firm that designed Linden Grove for the London Borough of Southwark. This is a small scheme of about 40 houses plus sheltered housing for old people, in a rather drab backwater of the borough. The design uses a pedestrian way down the centre of the long, narrow site. The 2 and 3 storey houses have their front doors on the pedestrian way and gardens on the other side, giving a high degree of privacy. The scheme is very intricate and carefully worked out, and is very well liked by the tenants.

Unlike Dawson's Heights, the site had little character of its own: however for Neyland it plays an important and almost mystical part.

There is so little to go on, in modern architecture, the site does become frightfully important, it's one of the only subjective non-measurable things. ... We try to get the building to respond and breathe with its surroundings.... To take an extreme, an appalling site can dictate the whole kind of [solution] ... where there are no views outwards you tend to work round the outside and look inwards, and create what you can in the middle, and vice versa. I mean, one of the reasons for Linden Grove was that there were some good trees round the edge of the site. So we tended to put it in the middle, looking outwards. That was one of the generating things.

Earlier in the interview, replying to a question on whether they used type plans developed in previous schemes, he said:

We don't think, as an absolutely central thing, that a type plan *should* be considered outside its site, I mean the whole point of good housing is the relationship between the unit and what's around it.

So the site was a generator; another objective from the outset was to build in low rise. But for this designer it was clearly difficult to separate out different factors in the design: the process and the product are seen in a holistic way. The various requirements are facets of a single problem, to be solved in an integrated way. The subjective is to be treasured. The analysis–synthesis model is clearly wrong here, but the fit with the generator–conjecture–analysis model is good. The site is described as a 'generating thing' without prompting or previous use of this term by the interviewer.

Possibly the site would be found to be the single most common generator in housing, if the design process was studied for a large number of schemes. For one more scheme in the sample the site played a role in generating the solution, but for the other three it was not very important in that they used solutions which had been at least partly worked out for other contexts.

The other scheme where site did play a part was Kedleston Walk, Bethnal Green, by Douglas Stephen and Partners. The scheme is of 58 units, for 2-, 4- and 6-person households, in an ingenious plan, again using a pedestrian way down the middle. To one side are 4-person maisonettes with small gardens, and to the other are 2-person flats with 6-person maisonettes above them. The roofs of the 4-person units serve as roof gardens for the larger households, reached by bridges across the central walk way. Douglas Stephen was interviewed; he was not the job architect but was involved with the evolution of the design. He is politically active and it could be said that a generating factor in the scheme was his awareness of the choice available to the council tenant, in comparison to the owner occupier. A major objective was to provide opportunities for people to live the way they want to, including opportunities resulting from being on or close to the ground. Another was the discovery of a local traditional form of working-class housing, using pedestrian streets.

I don't think of house plans at all at the beginning ... I think entirely of the site and of the restrictions, and there are not only spatial restrictions but also social restrictions on the site. Now (this site) was very difficult; it was expected of me that I would put up either two slab blocks or two tower blocks, and the first battle I had with the clients was actually to prove to them that you *could* do a low rise high density scheme. Now my argument was that the area was very deprived of public open space but possibly even more so of *private* open space. There was nowhere where anybody could be alone, or do any of the things that less deprived areas would be able to do, and it was on that particular level that I decided to try and do a sort of funny section, to spread the building as much as I could over the whole of the site.

Following this conjecture the designer had to analyse his proposal, to convince first himself and then the clients that it did in fact satisfy the problem.

I spent a lot of time wandering about [in Bethnal Green] and ... almost in the immediate vicinity of that particular site there was a particular Bethnal Green type of dwelling ... it was one of the bases for this site ... [He then described these dwellings, with diagrams, and how the pedestrian street formed a communal recreation space]. I so enjoyed this that I consciously tried, if not to *recreate* it, certainly not to let it entirely disappear, so when the configuration of the central street came about I did have this somewhat in mind.

This desire to maintain traditional patterns appears again as a strong element in Darbourne and Darke's Marquess Road scheme. The use of relatively low rise solutions at very high densities, pioneered in their earlier Lillington Street scheme, had originally been for visual rather than social reasons, to give a traditional townscape of streets and squares rather than the then current 'towers in a park' aesthetic. This aim was extended in Marquess Road to include the social objective of providing 'an atmosphere of house-dwelling for everyone', with a small private garden for family dwellings 'in a traditional relationship to a communal space'. A height of four to five storeys over most of the site was to keep the traditional London scale and to allow the church to retain its role as visual focus. The dwelling type plans were taken over virtually unchanged from stage III of Lillington Street, with the addition of a new two-storey type which was to be used beside the higher blocks, 'like a mews street' as the designers put it, again relating their forms to traditional patterns.

The designers did subject their scheme to rigorous testing against various constraints, including a feasibility study of rehabilitation of the existing buildings as an alternative to new building. But the aim of allowing traditional visual and social patterns to be retained is clearly a self-imposed constraint that arises from a value judgment. It therefore takes its place as an archetypal primary generator.

The next scheme to be considered is Pollards Hill, Merton. The interview was with Richard MacCormac, who has written about the scheme (MacCormac, 1971) and its debt to the land-use theories of March and Martin (see Martin and March, 1972). MacCormac was in fact one of a team of designers then with the London Borough of Merton. The site is a large one, about 41 acres, developed at 116 persons to the acre with an almost continuous three-storey terrace of housing that is folded round the outside of the site like a Greek key pattern, or 'a sort of intestinal geometry' in MacCormac's words. A series of 'P'-shaped courtyards are formed to both sides of the block, used on the outside for parking and on the inside as grassed play areas, giving access to the large central green space.

This was a particularly clear case of the design team learning about the problem by trying to produce a solution. The design team had been exploring the problem for some time, and had rejected a solution using maisonettes, before they articulated and clarified the criteria which they had been using unconsciously in rejecting the maisonette solution. When made explicit, these criteria could be used to produce a solution acceptable to the design team. A complicating factor was that they were attempting to design a housing system for use by a small consortium of local authorities for densities up to 150 ppa. This involved not only political problems but difficulties with cost when building to densities lower than 150.

The high price, the difficulties of the consortium, and our increasing awareness of the difficulties of building buildings that put families with children off the ground, all ... converged to make us feel that it was necessary to have a radical rethink.... So we then brought into focus a series of criteria which had been quite implicit in our criteria for the high buildings, the stacked buildings, but were all clearly based on what you can do with a house on the ground.... They were, to provide private open space, bigger than a balcony—obviously you can't do that economically unless you do it on the ground,—to connect that private open space to a public open space, of a kind that's associated with a limited group of people ... who can feel some sort ... of responsibility for the space, and will police it, ... and I can't quite see how you can do it in stack buildings. Then we thought that access, generally, was much better to a house than to flats, it's more convenient and there aren't indefensible (to use Oscar Newman's term) sorts of space.

Once we'd flipped from a stack dwelling to houses on the ground we assumed a terrace would be the best way of doing it ... and the whole exercise, formally speaking, was to find a way of making a terrace continuous so that you can use space in the most efficient way.

In other words, once the criteria were made explicit they could act as primary generators: the terrace concept was established and the analysis following this conjecture could concentrate on achieving house plans, site layout, and detailed design which best met their criteria.

Later, when asked about the source of the brief, MacCormac said:

a brief comes about through essentially, an ongoing relationship between what is possible in architecture and what you want to do. And everything you do modifies your idea of what is possible ... you can't start with a brief and (then) design, you have to start designing and briefing simultaneously, because these two activities are completely interrelated.

This is a particularly significant statement coming as it does from a designer trained during the sixties at two of the schools which emphasized 'design methods'. He was presumably taught to be aware of and in control of his own design process. In spite of such a training he finds the 'analysis-synthesis' methods unworkable and instead, as this paper has been suggesting, explores what is possible by making a conjecture at a solution.

CONCLUSIONS

The object of this paper has been to augment our understanding of the process of design as practised by architects. It is not the author's intention to prescribe a single 'correct' procedure, to criticise methods which differ from the model that has been proposed, or to provide a recipe for 'good' design. This was typically the intention during the early phase of design methods research, which has been briefly described above. The method advocated in the sixties was one of analysis followed by synthesis, or an elaborated version of this. In particular, the requirements and their mutual implications were to be thoroughly studied before any attempt was made to reach a design solution.

It has been suggested in this paper that designers do *not* start with a full and explicit list of factors to be considered, with performance limits predetermined where possible. Rather they have to find a way of reducing the variety of potential solutions to the as yet imperfectly understood problem, to a small class of solutions that is cognitively manageable. To do this, they fix on a particular objective or small group of objectives, usually strongly valued and self-imposed, for reasons that rest on their subjective judgment rather than being reached by a process of logic. These major aims, called here *primary generators*, then give rise to a proposed solution or conjecture, which makes it possible to clarify the detailed requirements as the conjecture is tested to see how far they can be met.

The evidence of the interviews with architects has supported the supposition that this is the way that many architects do in fact design. They use phrases such as 'the site ... was the main starting-off point', 'that was one of the generating things', and 'we had three major simplistic aims at the outset, before we started thinking about the design in detail'. It is clear in most cases that the design concept was arrived at before the requirements had been worked out in detail, and necessarily so, since these requirements could only become operational in the context of a particular solution. There do not appear to be any cases in the sample where the requirements and their interrelationships were analysed in detail in advance of any conjectured solution.

Thus the analysis–synthesis model would seem to be refuted as a method which can readily be used in practice. The generator–conjecture–analysis model has a closer fit with the evidence presented here. Of course this model should be subjected to further testing, and future research will, it is hoped, be able to elaborate the model, perhaps to differentiate between the methods of different designers, and to explore further the mental constructs that give rise to the primary generators. One of the shortcomings of the early phase of the design methods research was that it concentrated on *design morphology*, a sequence of boxes bearing particular labels, rather than the way particular designers filled the boxes with concepts, and the source of the designers' concepts. The author feels that the most interesting direction for design research to take now is to find further ways of 'looking inside the designer's head', of exploring subjectivity. The denial of the value of the subjective and the hope that the building would 'design itself' now seem to be products of a scientistic rather than a scientific way of thinking. The image of the user implied by this attitude was a mechanistic one, an anthropometric mannikin with certain environmental needs but no emotional responses. The users' reactions, often literally violent, to buildings designed with this image of man have shown that such architects and architecture are hated by the public.

A revaluation of subjectivity in design can lead to a revaluation of the subjective responses of the user, and hopefully to a more responsive architecture. Such an architecture will reflect the diversity and anarchy of human life, just as research on design methods should reflect the diversity in approaches to design. It is not necessary to prescribe one correct procedure to cover all cases; the trend in many branches of knowledge is away from the concept of a single reigning theory. Feyerabend (1975) has recently pleaded for a less authoritarian view of knowledge:

anarchism is not only *possible*, it is *necessary* both for the internal progress of science and for the development of our culture as a whole. And Reason at last joins all those other abstract monsters such as Obligation, Duty, Morality, Truth and their more concrete predecessors, the Gods, which were once used to intimidate man and restrict his free and happy development: it withers away.

Acknowledgments

The author wishes to thank Dr Bryan Lawson for his advice, and the RIBA for financial support.

REFERENCES

Alexander, C. (1963), 'The determination of components for an Indian village', in Jones, J. Christopher, and Thornley, D. G. (eds) *Conference on Design Methods*, Pergamon Press.

Archer, L. B. (1969), 'The structure of the design process', in Broadbent, G., and Ward, A. (eds) *Design Methods in Architecture*, Lund Humphries.

Aron, R. (1964), *German Sociology*, The Free Press of Glencoe.

Broadbent, G. (1973), *Design in Architecture*, John Wiley, Chichester.

Chermayeff, S. and Alexander, C. (1963), *Community and Privacy*, Doubleday.

Eastman, Charles M. (1970), 'On the analysis of intuitive design processes', in Moore, G. T. (ed.) *Emerging Methods in Environmental Design and Planning*, MIT Press, Cambridge, Mass.

Feyerabend, P. (1975), *Against Method*, New Left Books.

Hanson, K. (1969), 'Design from linked requirements in a housing problem', in Broadbent, G., and Ward, A. (eds) *Design Methods in Architecture*, Lund Humphries.

Hillier, W., Musgrove, J., and O'Sullivan, P. (1972), 'Knowledge and design', in Mitchell, W. J. (ed.) *Environmental Design: Research and Practice*, University of California.

Kuhn, T. (1962), *The Structure of Scientific Revolutions*, University of Chicago Press.

Lawson, B.R. (1972), 'Problem solving in architectural design', Doctoral Thesis, University of Aston in Birmingham.

MacCormac, R. (1971), 'The evolution of the design', *Archit. Des.* (October), 617.

Martin, J.L. and March, L. (eds) (1972), *Urban Space and Structures*, Cambridge University Press.

Maver, T. W. (1970), 'Appraisal in the building design process', in Moore, G. T. (ed.) *Emerging Methods in Environmental Design and Planning*, MIT Press, Cambridge, Mass.

Wehrli, R. (1968), 'Open-ended problem solving in design', Doctoral thesis, University of Utah.

Whitehead, B. and Eldars, M.Z. (1964), 'An approach to the optimal layout of the single storey buildings', *Archit. J.*, **139**, 1373.

3.2 An Exploration of the Design Process

Omer Akin

INTRODUCTION

The purpose of this study is to explore the process of design as it is manifested in the behaviours of human designers. Although design is basically a 'human' process by definition, it is necessary to differentiate it from *machine-design*, or computer-aided design, and *design-methods*, or rational design tools for human designers. As opposed to these the kind of design that is the topic of this study has been called *intuitive-design*, in the past (Eastman, 1970).

The payoffs expected from a venture like this have to do with other forms of design as well. First of all, it is necessary to understand intuitive-design to predict the performance criteria useful in developing appropriate tools for machine-design or design-methods. In the past, the biggest road block to the widespread use of design tools in architectural offices has been the incompatibility of these non-intuitive tools with those of intuitive-design. Neither the objectives nor the component parts of the intuitive-design process were understood well enough to enable a successful meshing of the intuitive and non-intuitive processes of design. Hence a better understanding of the former processes will inevitably lead to the successful incorporation of the two.

A second payoff can be expected in the area of architectural education. Better understanding of the intuitive-design processes is essential if students of architecture are to be freed from the esoteric and inefficient methods of training used in the traditional design studio in lieu of explicit instruction and

Originally published in *Design Methods and Theories,* **13** (3/4) (1979), 115–19.
Reproduced by permission of The Design Methods Group.

procedural feedback. If we expect better results from the criticism technique employed in design studios, the process of arriving at a 'bad' design should be under criticism more so than the 'bad' product itself.

Finally, making the design process explicit will no doubt influence the way professional designers design, formulate problems, and evaluate designed products. The mysticism attached to the process of design has been used long enough to justify 'bad' designs, even in cases of absolute well intent.

Based on these three sources of motivation, this study aims to accomplish two specific goals:

(1) Disaggregate the process of design: most of the time understanding of a phenomenon means discovering its parts and their relations under different circumstances. By exploring the components of the intuitive-design process I hope to provide a basis for interfacing it with non-intuitive processes as well as training students consciously to interact with their design processes.
(2) Compare intuitive and non-intuitive processes: by evaluating the efficiency of alternative processes it will be possible to compile new aggregate processes that may develop into 'better' forms of intuitive-design in the long run.

In the major portion of the remainder of this paper I shall deal with the former goal. First, a method of examination of the behaviours of the designer will be developed. Next, application of the method and its findings will be reported. Finally, the properties of the intuitive-design process developed in the findings section will be contrasted against the properties of the non-intuitive methods. Although a broad scope of knowledge is necessary to explain adequately many of the premises and findings of this study, it is essential to keep the discussion of other works to a minimum in order to prevent it from turning into a literature review paper. Most of what is left unstated can be found in the sources referenced in the text.

METHODS OF EXPLORATION

Human behaviour has been measured and calibrated by means of three basic techniques; (1) reaction-time studies, (2) eye-fixation studies, and (3) protocol analysis studies (Newell and Simon, 1972). This basically dictates a selection between two alternative experimental attitudes. One is to construct a well-defined and controlled experiment where various components of subjects' behaviours can be parametrized and measured using reaction-time and/or eye-fixation data. The second is to collect less quantifiable and informal data that reflect a broader scope for the design behaviour that is being examined, i.e. protocol data.

The former alternative requires that the behaviour under examination is understood *a priori*, to the extent that it can be decomposed into component parts and/or that concrete hypotheses about it exist. Given that this background knowledge is essentially lacking in design and the task of this paper is to come up with those concrete hypotheses in the first place, the latter method is more suitable here. Protocol analysis is especially suitable for my purposes, for several reasons:

(1) it provides a context in which a task environment, such as design, can be explored in its entirety, i.e. in aggregated format;
(2) *a priori* experimental hypotheses about the task environment are not essential;
(3) it can provide rich data in the form of verbal protocols as well as reaction times and/or visual attention cues.

Protocol analysis

Protocol analysis is a technique devised to infer the information processing mechanisms underlying human problem solving behaviour (Newell, 1968). A protocol is the recorded behaviour of the problem-solver. It is usually in the form of traces or recordings of the overt behaviours of problem-solvers, such as notes, video or audio recordings of behaviours, sketches, etc.

All problem-solving behaviour is assumed to consist of transforming a given state of information about the problem into another state such that the second state is closer to containing the information that describes a solution to the problem than the first one. The act of transforming a problem state into another one is called an operation. Given this taxonomy a protocol starts with an initial information state (or the *problem description*) followed by a sequence of many intermediary states before reaching a final (or *solution*) *state*. Each state can be obtained by the application of an operation (or *operator*) to the previous state.

Imagine a network of *states* (or beads) and *operators* (threads that join two successive beads) such that loose-ended sequences of beads can also be attached to form this network. This representation is called a problem behaviour graph (PBG). In this representation the task of solving a problem corresponds to finding the right sequence of operators (or a path) that transforms the problem description state into the solution state. The 'loose-end' paths in PBG represent those sequences of operations that correspond to unsuccessful attempts at reaching a solution, i.e. dead-ends.

Since the PBG provides: (1) a catalogue of operations that can be applied in a task environment, (2) the circumstances under which such applications are made, and (3) the paths developed during the 'search' for a solution, it is very useful in formalizing process models for explaining and simulating the task environment in question. Any flow-diagram composed of the operations found in a PBG, and that can generate behaviours essentially identical to the data found in the protocol, is a candidate for such a process model. The validity and reliability of the model can be checked by comparing its 'behaviour' against behaviours obtained at different times and/or from different subjects. Often the behaviour of a process model is much too complex to simulate manually. Hence, computer simulations are the norm in evaluating and verifying process models defined after the human subject.

Design problem solving

Architectural design is a form of problem-solving. Many researchers have pointed to the pecularities of design (ill-definedness) compared to other forms of well-studied problem-solving tasks such as theorem-proving, chess, semantic processes, etc. (Reitman, 1964; Newell, 1970). On the other hand it has been shown repeatedly that design can be adequately represented using the taxonomy of PBGs and information-processing models. (Foz, 1973; Krauss and Myer, 1970; Eastman, 1970). The basic difficulties in these studies can be attributed to inherent problems with the protocol analysis techniques more so than difficulties arising from the ill-defined nature of design.

Critics of protocol analysis techniques assert that:

(1) often subjects are asked to verbalize their behaviours during protocol experiments—subjects introspecting about their actions have been shown to be in error, in many studies;
(2) the thought process, being much faster than motor responses, cannot be fully reflected through the motor behaviours of subjects;
(3) due to the magnitude of the analysis required to understand the average protocol data, very small numbers of subjects are used in each experiment;
(4) there are usually gaps (or periods of silence) found in most protocols, which cannot be accounted for by proposing a complete lack of cognitive activity.

The first and the third criticisms are totally unfounded. Verbalization during the act of problem-solving is no different from reacting to a stimulus in a reaction time experiment. Most, if not all, verbalizations are not *a posteriori* introspections but are genuine reflections of problem-solving behaviour. The small size of samples used is greatly offset, on the other hand, by the hundreds or thousands of observations found in the protocol of each subject. Conclusions reached at the end are generalizations about the consistencies within each problem-solver rather than between many individuals.

The problem of 'missing data', however, is doubly significant in design protocols. A lot of visual and implicit processing cannot be expressed in words with ease; so subjects seem to ignore reporting such experiences fully. Furthermore, due to the specification of most of the design goals in the latter stages, most 'dead-end' paths are discovered only later, and are usually not reported by subjects. Hence, careful consideration must be given accurately to interpolate 'missing data' in the analysis of protocols.

PROTOCOL ANALYSIS IN DESIGN

Two aspects of the protocol analysis procedures outlined above have to be altered before they can be applied to design tasks. These are the problem behaviour graph (PBG) and the computer simulation phase in which the models developed are tested. Design is a very complex information-processing task. Our present programming and hardware capabilities are not up to par for simulating the semantic and visual processes that constitute design. Hence the computer simulation phase will not be attempted here.

The PBG found in design protocols usually contain large gaps due to the 'missing data' problems cited above. Consequently, design PBGs appear to be much less 'bushy' than most other PBGs. Instead of trying to compensate for this by second-guessing the subjects the PBG representation will not be used here. Instead the notion of 'Plans' (Miller *et al.*, 1960) will be used to codify the various design operations used in design.

Plans are hierarchical processes that control the order in which a sequence of operations is to be performed. To fulfil their needs and intentions humans plan and execute series of actions. Miller *et al.* propose that this provides the essence of all human behaviour in pursuing goals, or solving problems. Each Plan takes an information state as input and produces another state, in much the same way that operators in PBGs do. Plans can be executed simultaneously, such as 'watching TV' and 'eating'. Plans also come in different sizes and with different scopes. Higher level Plans, or Strategies, consist of many Plans; and Plans themselves are made up of many lower-level Plans, or Tactics. Many-level nestings of Plans can be formed as long as they are useful in representing human behaviour.

A useful way of disaggregating complex behaviour is to identify Plan hierarchies and show the conditions of *input* and *output* related to each Plan. This not only yields a chaining of actions in much the same way the PBG does, but also shows the complex relationships of simultaneously executed Plans.

The task of identifying Plans and demonstrating their experimental reliability is a difficult one. There is no precedence or scientific theory to facilitate this task. Based on general knowledge, two criteria can be proposed as measures of reliability in design protocol analysis: plausibility and consistency.

Plausibility is necessary in making 'better' inferences in identifying design Plans. The basic logical tool in infering design-Plans from protocols is induction. The only insurance one has in correctly infering a *rule* (Plan) from a given *case* (input information state) and a *result* (output information state) is knowledge of what is plausible based on past experience.

Let me illustrate this point with an example. Suppose the subject requests to have a ruler (input state), and after obtaining the ruler he sets out to measure the drawing he has. We can propose that the subject's goal was to measure something in the drawing. He developed a Plan consisting of: getting a ruler, placing it against the drawing, and measuring it. We can alternatively propose that the subject's initial goal was to scratch his back. So he developed a Plan which starts with getting the ruler. Upon obtaining the ruler, however, he is reminded of the thing he needs to measure in the drawing. So he forgets about his itching back and starts to execute the Plan for measuring the drawing.

Notice that in each case the input and output to the inferred Plans are constant. Only the alternative Plans vary. It is also obvious that the former alternative is the plausible one. Although most alternatives with fundamentally different compositions of Plans are not likely to be equally plausible, some are. In such cases the criteria of consistency can be used.

Consistency criteria are necessary for providing an independent test for measuring reliability. It requires that the proposed Plans are not inconsistent with (1) other Plans identified in the present task, (2) Tactics and Strategies compiled in other studies of human information processing, and (3) Plans identified at different times or with different subjects.

EMPIRICAL FINDINGS

In accordance with the principles set forth above, a protocol analysis of design behaviour was carried out to understand the nature of the architectural design process. A design problem of moderate size was given to a professional designer under laboratory conditions. The problem, a single-person dwelling, was generated in a previous experiment. The problem statement and the site plan are provided in Table 1 and Figure 1 respectively.

The subject developed a sketch design made up of two plans, one section and one axonometric in approximately four hours. The sketch plan is shown in Figure 2. The experiment in its

Table 1 Problem description document: problem brief: brief for a Single-Person Dwelling (based on an interview with the client)

The problem is to design a garage with occupancy for one person over it or around it in some way. The site is on a lot currently occupied by a house and its garage. The site of the existing garage is donated by the owner of the house to the Association for the Blind. The new building will replace the existing garage. The area to the right of the garage is the private back yard of the house which will continue to be used by its present owners. The new garage has to handle two full-size cars and 600 cubic feet of storage space for the use of the owner of the house.

The purpose of the donation is to provide a permanent place of residence for guests and teachers related to the School for the Blind on Craik Street. Different persons are expected to occupy the residence over a long period of time. All users are assumed to be blind and they are most likely to be teachers at the School for the Blind.

The objective of the design is to provide a rich personal environment for the blind user of the residence. Currently we have access to the person who will be occupying the building as soon as it is completed. Possibly in three years another person will move in her place.

The user will live here, eat here, and sleep here. She will use it as a base of operations for her teaching activity in the School for the Blind. She is well educated; has braille machines and other technical equipment for recordings etc. She teaches blind persons of all age groups.

Since she has been blind for most of her life the sensory experiences of blind persons and non-visual aesthetic considerations must be regarded in designing this building. Owner of the house has also set up a trust fund for the building and its maintenance. The total fund is $60,000, as an endowment.

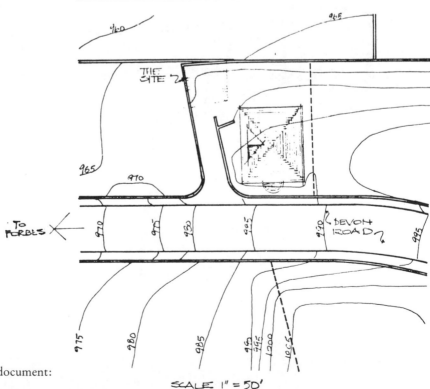

Figure 1
Problem description document:
site-plan.

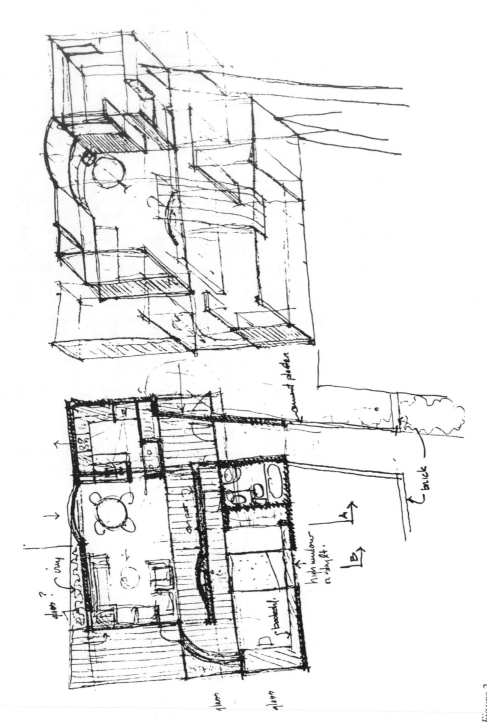

Figure 2
First-floor sketch plan and
axonometric.

entirety was recorded on video-tape. The audio part of the tape was transcribed and used in the analysis in conjunction with the notes and sketches of the designer. Figure 3 and Table 2 contain samples of the subject's sketches and the protocol, respectively.

Analysis

After getting acquainted with the contents of the video recordings, a preliminary analysis was conducted. Many patterns of actions that suggest Plan-like behaviours were identified. This accounted for about 90 per cent of the protocol. In this first pass about 30 plausible Plans were discovered.

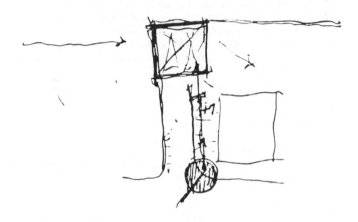

Figure 3
A sample sketch from the protocol.

Careful re-examination revealed that there were similarities of input and/or output between most of these Plans. By translating some of the constant parameters used in describing these Plans into variable parameters the total number of Plans was reduced to eight. Miller *et al.* call Plans with variable parameters, Schemata.

In this second pass considerably greater consistency among inputs and outputs of successively activated Schemata were observed. The consistency of these Schemata with respect to other task environments will be tested by examining their relationships to Strategies and Tactics discovered in these task environments. But we must first define the eight Schemata that account for the protocol; instantiation, generalization, enquiry, inference, representation, goal-definition, specification, and integration.

Three things are necessary and sufficient in defining Schemata; *input, output,* and the body of *actions* needed to convert the input into the output format. The input is the information state which is necessary for activating Schemata. The output is the information state that is generated by Schemata. *Actions* are pre-compiled and goal-driven behaviour sequences. They can exhibit variable behaviour, hence produce variable output, as a

Table 2   Sample protocol (page eight of transcript)

line	
0	(subject has just considered placing the residence he is designing under the existing garage)
1	probably the expense would be considerable to do that
2	it would solve the entry problem
3	it would also save some money from building the structure, the physical outside walls
4	we probably will have to rebuild the structure of the garage
5	which might defray some of that savings
6	so I have an idea to do some thing
7	(experimenter shows slide 9) That is the corner of the house
8	(experimenter shows slide 10) I think that is the last slide
9	would it be possible to go back and look at the slides later?
10	(experimenter) Sure, any time
11	so there is quite a bit of flexibility in the house
12	there is a number of different approaches we could take
13	it's going to be one person living here?
14	and that one person would probably need a living area,
15	dining area, kitchen, bath, bedroom and perhaps study for the equipment
16	so it is essentially a one bedroom unit
17	which could be put in easily,
18	not easily but in 625 sq. ft.
19	o.k. I think what I would like to do is to sketch some organizational
20	(experimenter) I will start the video
21	(experimenter) and you can use some of the paper and drafting equipment here
22	(experimenter) just use whatever you're most comfortable with
23	start out with ... (mumble) (starts drawing Figure 3)
24	okay ... we have ... (draws south wall of the lot)
25	we have a site ... that is (draws west wall of lot)
26	existing garage ... that (draws outline of existing garage)
27	walk through right here (draws curved retaining wall)
28	there are some trees here ... (draws trees south of driveway)
29	I think the first thing I'm going to try is
30	I think ... I found often that the first ideas are often the best
31	after examining several ideas are the best
32	so I'll first experiment with the idea of trying to enter right here (draws line indicating entry from the corner)
33	leave the existing garage where it is
34	take the roof off (re-draws the garage outline)
35	and add the building over
36	problems being right at that point (points to curved corner)
37	(experimenter) would you elaborate that?

function of the parameter values of the inputs. This is similar to the behaviours one can expect from conditional statements of the kind: if X then Y, where X represents the value of the input parameter and Y the kind of action initiated in the Schemata. All eight Schemata are described below using the input–action–output sequence.

Schema A: Instantiation

Input: An external symbol is encountered, such as a word or a visual sign in the problem documents. An internal symbol is encountered, such as a concept or image originating in the designer's mind.

Action: If the symbol is recognized as an instance of a general class of symbols, such as chair, chimney, access, etc., then it is candidate for instantiation. Otherwise the value in attempting to recognize the symbol must be assessed. If the symbol is interesting enough to warrant exploration, then more sources of information will be examined in order to recognize it. Otherwise the symbol will be ignored until new interest in it comes about. When a symbol becomes a candidate for instantiation it is checked against symbols that are part of the problem representation. If the symbol recognized is novel it gets instantiated as a member of a class of symbols in the problem representation.

Output: A symbol is saved, or instantiated as a part of the problem representation. More sources of information are searched. Input is ignored.

Schema B: Generalization

Input: New symbol(s) are encountered such that with other related symbols they form a set with recognizable attributes, such as clear, dominant, central-dome house, etc.

Action: If the set of symbols being attended to constitute a supra-symbol, i.e. a symbol ('seating') composed of many other symbols ('chairs'), then the attribute recognized can be associated to the supra-symbol as well. Otherwise the input is ignored.

Output: An attribute is associated to a supra-symbol. Input is ignored.

Schema C: Enquiry

Input: There is a need for searching new information sources (from Schema A). Symbol(s) newly instantiated are inconsistent with other instantiated schemata. New symbol(s) may clarify the 'meaning' of previously ignored symbol(s).

Action: If there is a plausible explanation, or a hypothesis, for the inconsistency observed, then examine *external sources* such as problem documents, previously known publications, etc.; *internal sources* such as mental simulation of the real world, or *ask* the experimenter about the hypothesized explanation. If no specific explanation can be hypothesized ask a general question related to the symbol(s) in conflict or ignore input.

Output: Instantiate new symbols or reinstantiate old symbols to eliminate the initial inconsistency, based on new information gathered from external or internal sources. Ask experimenter a question. Ignore input.

Schema D: Inference

Input: There is a need for generating new information from internal sources (from Schema A). New symbol relations are formed (from Schema B). New symbol relations are hypothesized to exist (from Schemas F and G).

Action: If there is a desired direction for inference-making, such as given hypothesized relations, then carry out goal-driven inferences. To make goal-directed inferences find a set of plausible inferences in order to transform any subset of the problem representation into the desired information state. To make spontaneous inferences, such as when there are no hypotheses or goals proposed, start from a known information state and transform it until an interesting state is reached.

Output: New symbol relation(s) are discovered or hypothesized.

Schema E: Representation

Input: Information transformations with visual imagery are to be made (from Schema D). New instantiations have to be stored externally (from Schema A).

Action: If there is a representational format suitable to accept the symbol(s) to be represented, such as section, site plan, bubble diagram, etc., then sketch, build, or note symbol(s) externally. If no compatible representation is available then either form a new, suitable representation or transform the symbol into a format suitable for the representations at hand.

Output: Make external representation, such as sketch, drawing, model, notes, bubble diagram, report, computer program, etc.

Schema F: Goal-Definition

Input: The current solution developed is not satisfactory (from Schemata G and H). There are more than one goal or sub-goal candidates to be pursued.

Action: Based on pre-compiled, or archetypal, solutions and performance criteria developed through experience define short-term attributes to be built into the current problem state. Specify Plans, or Schemata, necessary to accomplish these transformations (or goals). If more than one such goal is available select the most efficient or promising one. If goal-directed Plans can be activated only after prerequisite Plans are used to bring about the necessary changes in the current information state then set up these prerequisites as sub-goals. Return to goals after satisfying sub-goals.

Output: Define goals, or sub-goals. (A goal or sub-goal can activate any Schema.)

Schema G: Specification

Input: The current state of the solution is not specific enough for the purpose of the design task, such as idea development, feasibility, construction, etc.

Action: Find a transformation which will make the current solution more specific, such as a pre-compiled solution, an analogous solution, a generic solution, etc. Use inference rules to reconcile incompatibilities in the new representation. If explicit transformations are not possible or feasible at the time, use previous experience to assume that certain aspects of the current solution can be further 'specified'.

Output: A partial solution, or specification of a certain aspect of the current solution.

Schema H: Integration

Input: New partial solution(s) are generated.

Action: If the new solution(s) are inconsistent with any parts of the present problem state then integrate them. Otherwise, instantiate the newly developed information. To integrate new and existing solutions determine the one that is more dominant or significant. If they are equally dominant, then save both as alternative solutions and consider a new goal for exploring the potentials of the alternatives. If there is one solution dominant over other(s) adapt weak solution(s) to the dominant one either by modifying some attributes of the weak solution or by redefining problem constraints in such a way to accommodate all solutions.

Output: Further specification of the current solution state. Need for a new goal or sub-goal.

Focus of Attention in Design

The specific context within which these Schemata are applied by the subject can best be described through the architectural design issues he attends to throughout the protocol. For example, the first things attended to are geographic location of site, its orientation, the kind of design expected, relations

Table 3
Analysis of protocol

Hierarchy of strategies			Focus of attention	Plan Code	Line number	Page number
Set-up design context: acquisition and assimilation	Activate conceptual knowledge (pattern recognition)	Identify task environment	Geographic locale and site	A	1	1
			Orientation	A	4	
			Unusual symbol	C	5	
			Task type	C	6	
			Site element relations	C	7	
			:			
		Identify parameters of task environment	User and his properties	E	1	2
			Site relations	H	3	
			Site user relations	D	3	
			Access issue	F	13	
			Programming issues	:		
			Budget			
	Activate visual knowledge (abstraction)	Identify visual parameters of task environment	House image	A	3	5
			Access image	B	7–8	
			Corner entry problem	C	8	
			Scope of task	D	10	
			Resolution of scope of task	:		
			'Fortress' image problem			
			Resolution of the fortress			
			Access alternatives			
			Spatial programming	C	13	8
			Feasibility of project	D	14	
			Set up visual representation	G	14–15	
			Assess alternatives	B	16	
				:		
		Set-up design representation	Topology of site			
			Grid systems on site			
			Access parameters			
Search cycle: develop design goals	Develop partial solution		Entrance	G	18	10
			Organization goal	F	21	
			Entry solution	H	22	
				F	23	
				:		
	Partial-solution		Living area			
	Sub-goal		Orientation	D	3–8	11
	Sub-goal		Blindness of user	H	9	
				F	10	
				:		
	Partial-solution		Public areas			
	Idea		Overall progress			

between site elements, and so on. Column 4 of Table 3 contains a complete list of all issues attended to by the subject in the first 18 pages of the 48-page transcript. The fifth column of Table 3 indicates the Schemata used in exploring some of these issues. The last two columns in Table 3 provide the line and page numbers of the transcript from which this information is extracted.

Table 3 cont.

	Hierarchy of strategies			Focus of attention	Plan Code	Line number	Page number
Search for design solution: hill-climbing	Search cycle: develop design goals		Develop partial-solution	Private areas	G	24	
				Semi-private areas	G	25	
				Evaluate solutions proposed	⋮		
			Partial-solution	All areas			
			Develop partial solution	Mechanical room	G	9	12
				Gas on site	G	10	
				Electricity on site	G	11–12	
				Mechanical room	⋮		
			Partial-solution	Wet and dry service areas			
			Idea	Evaluate relation to house	F	40	13
					H	41–43	
				Overall progress	H	1	14
					⋮		
	Search cycle: develop design goals		Idea	Core idea from P Johnson			
			Evaluation function	Evaluate progress			
			Develop partial-solution	Entrance	H	25–26	
				Garage	G	27	
				Entrance	⋮		
				Assess progress			
			Develop partial-solution	Entrance	H	4	15
				All areas	G	5–6	
				Assessment: Orientation	H	9	
			Partial-solution	Partition users' territories	G	10–13	
				Assess solution	E	14	
					⋮		
			Representation	Inter-house relations			
			Partial-solution	Entrance and dining room	G	10–11	16
					G	12–14	
					⋮		
			Idea	Partitioning and grid systems			
	Search cycle: develop design goals		Develop partial-solution	Assess partitions and grids	D	3	17
				Assess partition	H	31	
				Propose new partition	⋮		
			Partial-solution	Core and Bedroom	E	6	18
				Assess orientation	G	7	

The data in Table 3 indicate several general observations:

(1) No single issue occupies the subject's attention more than 10 minutes at a time.

(2) Initially, during the first 26 issues, attention is largely dependent on external stimulation, such as problem docu-

ments. Later on, however, control of attention seems to respond more closely to internal processing needs.

(3) There seems to be a correlation between the issues being attended to and the representations used. Visual representations, for instance, go best with solution specification, and so on.

(4) Attention also seems to depend on the amount of information gathered about certain issues at that point in time. For example, issues of organization do not come up until the overall context of the design is adequately understood. This suggests that there is an implicit hierarchy of the issues attended to in the protocol.

(5) Every once in a while, after an average of 14-minute intervals, attention is focused on the degree of success of the process to date and new goals are sought.

(6) All issues are picked up and dropped numerous times. Each time an issue is picked up a new solution or goal is developed. But these represent partial solutions that specify the design issue to a limited extent.

The combination of the focus-of-attention information with the Schemata applied to this information provides the complete story of the protocol. The former represents the context of the design activity while the latter its method. By examining what is being processed and how it is being processed, it was possible to disaggregate the design process into higher-level processes, or Strategies. The first three columns in Table 3 represent the major Strategies observed in the data in a hierarchical way.

Design strategies

The frequency distribution of the Schemata in Table 3 suggests two major strategies for design; *pre-sketching* and *sketching*. The characteristic Schemata used in the pre-sketching phase are Enquiry, Instantiation, and Inference. These three constitute 58 per cent of all Schemata observed in this phase. These are strictly useful in acquiring and assimilating external information. As soon as an internally assimilated representation of the problem context is formed a physical representation is developed and the design activity begins. This major strategy is labelled 'Set-up design context' in the first column of Table 3.

In the sketching, or search for design, phase the main Schemata used are Specification, Integration, and Enquiry. These three constitute about 59 per cent of all Schemata observed in this phase. The search process in general does not resemble forms of search valid for other task environments. While each search cycle is initiated to bring more specificity to limited aspects of the solution description, i.e. depth-first, the same cycles also serve to develop new goals for the design task, i.e. breadth-first.

It would not be simplistic to suggest some deterministic patterns for the Strategies used within the pre-sketching phase. First, problem documents are used to activate conceptual knowledge; later, visual knowledge is activated; and finally, physical representation of the context of the design problem is formed. It is reasonable to assume that this final stage is a prerequisite in developing physical solutions. The first two, on the other hand, serve the purpose of assembling the knowledge necessary to build this representation. First the semantic concepts about the problem are activated, such as, site, north, relations, etc., then visual imagery is used to illustrate and embellish the images related to these concepts. Several more familiar Strategies are observed as sub-strategies making up the above three, such as pattern recognition, matching forward processing and heuristics. (Akin, 1978.)

The sketching phase seems to have a single major control structure, i.e., hill-climbing. Hill-climbing is a search strategy in which search starts from any point in the problem space. All subsequent design operations try to move this point closer to the goal, or solution, state. Success or failure depends on whether the initial point selected is one that is potentially capable of leading to a satisfactory solution. Each search cycle in Table 3 starts from a different point of departure and tries to develop it into a solution, or partial solution.

Various Strategies that have been observed in other task contexts are parts of the strategies listed in column three of Table 3. Some unique forms of heuristic search are also seen in the protocol, such as (1) obvious-solution-first; (2) divide-and-conquer; (3) pre-compiled-solution; and (4) most-constrained-first. The more familiar forms of search Strategies observed are (1) hypothesize-and-test; (2) induction; (3) means–ends–analysis; (4) hill-climbing; and (5) pattern matching (Newell, 1968). Throughout the remainder of the protocol the patterns of Schemata used are similar to those in Table 3.

CONCLUSIONS

Although the findings provided above are tentative at best, it is clear that many plausible hypotheses can be formed about the nature of the design process which contradict some beliefs that have influenced the progress of design methods for some time. The first assumption I would like to take issue with is the analysis–synthesis–evaluation cycle that lies at the heart of almost all normative design methods proposed until now.

As we can see from the protocol, one of the unique aspects of design behaviour is the constant generation of new task goals and redefinition of task constraints. Hence 'analysis' is a part of virtually all phases of design. Similarly, 'synthesis' or solution development occurs as early as in the first page of the protocol. Not only is the compartmentalization of the design process into three rigid phases (i.e. analysis–synthesis–evaluation) untrue, but the tactics implied for each of these compartments are also unrealistic. All solutions do not arise from an analysis of all relevant aspects of the problem. Often a few cues in the environment are sufficient to evoke a pre-compiled solution in the mind of the designer. Actually, this is more the norm than a rational process of assembly of parts, as suggested by the term 'synthesis'.

Many rational methods of design violate a basic design criterion widely used by human designers, namely 'satisficing' solutions. In attempting to make the process of design 'scientific' it is generally believed that the 'best' design solution is the one that *optimizes* all factors that contribute to the design circumstance. However, it has been shown that this principle is not suitable for the design task environment. No quantifiable model is complex enough to represent the real-life complexities of the design process; and no computer system is efficient enough to optimize equations with so many degrees of freedom. This is why human designers 'satisfice' their solutions, i.e. pick a solution that satisfies an acceptable number of design criteria and performs better than a minimum value for each criterion, rather than trying to optimize them, i.e. estimate parameters to minimize or maximize the value of an objective function.

The process of design is primarily a top-down, breadth-first process. This implies that different aspects of the design activity are hierarchically dependent on one another. Hence, any method of design, whether it is a small aid to be plugged into the overall intuitive design process, or it is a complete design process itself, has to conform to the human parameters, if interface is anticipated at some level. Consequently, the study of the design process will yield new insights into the study of design methods by providing precedents for innovative techniques as well as providing new areas of research.

REFERENCES

Akin, O. (1978), 'How do architects design?' Proceedings of IFIP Working Conference held at Grenoble, France (March).

Eastman, C. (1970), 'On the analysis of intuitive design processes', in Moore, G. (ed.) *Emerging Methods in Environmental Design and Planning*, MIT Press, Cambridge, Mass.

Foz, A. (1973), 'Observations on design behaviour in the parti', Proceedings of the Design Activity Conference, London.

Krauss, R. I., and Myer, R. M. (1970), 'Design: A case history', in Moore, G. (ed.) *Emerging Methods in Environmental Design and Planning*, MIT Press, Cambridge, Mass.

Miller, G. A., Galanter, E., and Pribram, K. H. (1960), *Plans and the Structure of Behaviour*, Holt, New York.

Newell, A. (1968), 'On the analysis of human problem solving protocols' in Gardin, J. C. and Jaulin, B. (eds), *Calcul et Formalization dans les Sciences de L'Homme*, Paris.

Newell, A. (1970), 'Heuristic programming: ill-structured problems', in Arnofsky, J. (ed.), *Progress in Operations Research*, Wiley, New York.

Newell, A., and Simon, H. A. (1972), *Human Problem Solving*, Prentice Hall, Englewood Cliffs, N. J.

Reitman, W. R. (1964), 'Heuristic decision procedures, open constraints and structure of information processing problems', in Shelly, M. W. (ed.) *Human Judgements and Optimality*, Wiley, New York.

Simon, H. A. (1973), 'Structure of ill-structured problems', *Artificial Intelligence,* **4,** 181–201.

3.3 Cognitive Strategies in Architectural Design

Bryan R. Lawson

1. THE EXPERIMENTAL SITUATION AS A MODEL OF DESIGN

Whenever a group of architects come together to discuss the nature of architecture and design there is bound to be considerable disagreement. It is difficult to find a consensus definition of architecture except at a very general level. If an experimental study such as this is to make progress, we must retreat to this generalized and relatively abstract level of agreement. The architect's primary and central task is to produce a 'concrete' three-dimensional structure of space and form to accommodate an abstract structure of related human activities. The purpose of the experiments to be described is to understand how architects perceive the relations between variables in multidimensional design problems, and how they produce desired relations between the elements of their solutions.

If the results of an experiment are to be extrapolated to give conclusions about the real design situation, then it must model that aspect of the process with which it is concerned as accurately as possible. Most importantly here, the subject must produce a unique spatial configuration of elements so as to satisfy some abstract structure. The universe of possible alternative solutions must be so large as to make successive trial and error an uneconomic strategy but bounded so that any

Originally published in *Ergonomics*, **22** (1) (1979), 59–68. Reproduced by permission of Taylor and Francis Ltd.

solution the subject can produce may be predicted. Many experiments and tests require subjects to produce spatial configurations of elements, but they are not models of a true design process because they lack hidden structural rules.

The solution that our subject is to produce must be constrained by structural rules, which he must discover for himself. The task of discovering structural rules has been investigated by many psychological experiments, but there the discovery of the structure is the end, and not a means to an end as in a true design situation. In this experiment then, the discovery of the structure and production of the solution must be integrated and the subject must not be instructed to perceive them as separate tasks. It is also important that there is no one obvious 'right' answer to the problem, but rather it is left up to the subject (designer) to decide what is the best he can achieve.

If architects are to be compared with other groups then neither the discovery of the structure nor the production of solutions must require the subject to have specialist knowledge. Problems must not be posed that are likely to be differentially familiar to the various groups of subjects.

2.   THE EXPERIMENTAL TASK

The experimental material consists of four pairs of coloured blocks, and a rectangular plan mounted on a turntable. The two members of each pair of blocks are identically shaped but different to all other blocks. The top and bottom surfaces of one block in a pair are white, while those of the other are black. All remaining surfaces are either red or blue. The two blocks of a pair show different patterns of red and blue on the vertical surfaces. The blocks are numbered one to eight for identification purposes. The rectangular plan consists of a grid of three by four bays, each bay being 40 mm square. The blocks are also based on this dimensional module, and are illustrated in Figure 1.

In the experiment the subject was asked to arrange four of the blocks, one from each pair, on the plan so as to cover all twelve squares and with no blocks projecting. The blocks must be laid with the black or white surfaces uppermost, and the subject was asked to maximize the amount of either blue or red showing around the external vertical face; see Figure 2. The subject was told that not all possible combinations of blocks would be allowed each time, and a rule requiring certain blocks to be present would be fixed before each problem. The subject was not told what the rule was, but he was allowed to ask if a combination of blocks that he had assembled was acceptable or not. The subject would be given a simple 'yes' or 'no' response to this question with no detail as to why or why not in the manner of classical concept attainment experiments. The subject was asked to reach the best solution that he thought possible by asking as few of these questions as he could. The

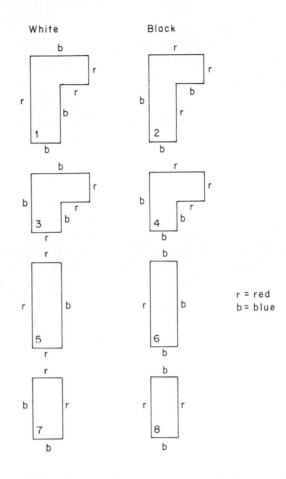

Figure 1
Plan view of blocks.

subject himself decided when that best solution had been achieved. There was no time limit.

The subject was thus not instructed to discover the rule, but rather to produce a solution, as in a real design situation. Thus

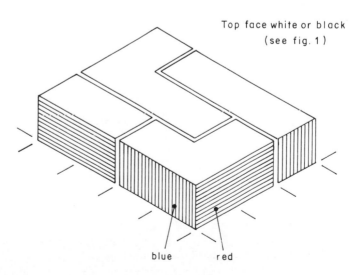

Figure 2
Typical arrangement of blocks.

the subject may produce a solution either as a result of discovering the rule or in complete ignorance of the rule. Three different types of rule have been used in the experiment reported here. They are:

Affirmation—one block must be present   (A)
Conjunction—two blocks must both be present (A and B)
Disjunction—either or both of two blocks
            must be present   (A or B)

The subject knew that each rule will be in one of these three formats. The letters 'A' and 'B' can refer to any of the eight blocks, except that A and B cannot belong to the same pair of blocks. There are therefore eight possible affirmations and 24 possible conjunctions or disjunctions, making a total of 56 possible rules. Since in each case the subject could be asked to maximize either red or blue, this makes a grand total of 112 possible problems.

It can be seen that this experiment has some affinity with a traditional concept attainment experiment of the type made famous by Bruner *et al.* (1956). In such experiments, however, the subject is asked only to discover structure not to produce it as well, and for this reason they are inadequate for the purpose of investigating the design process. (Although Lewis (1963) did use one for just that!) However the parallels are useful and Lawson (1969) demonstrates that in an array of four two-step variables the conjunctive and disjunctive rules are complementary, and objectively require identical amounts of information for their solution. However subjects showed much more difficulty with disjunctive rules; a finding suggested by the results of many previous workers. Indeed several workers have shown that subjects tend to perceive disjunctions as conjunctions or affirmations. Figure 3 shows that with the array used in this experiment there are eight acceptable solutions when the rule is an affirmation, four when a conjunction, and twelve when a disjunction. Thus in this model of the design situation, it is important to identify disjunctions

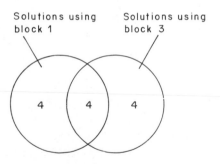

Figure 3
Number of solutions for different problem types.

since they give a greater range of possible solutions and therefore a greater chance of maximizing the area of required colour.

3. EXPERIMENTAL DESIGN

The experiment was piloted using final-year students of architecture as subjects who were told they were helping to develop the experiment and were thus encouraged to talk about what they were doing and their reactions to the problem, which were tape-recorded. The experimenter sat opposite the subject, set the problems, and gave information as needed.

In the main experiment the experimenter was replaced by a Digital Equipment Corporation PDP9 computer used on-line. No record was made of subjects' comments.

The use of an on-line computer connected to the laboratory by a remote teletype terminal left the subject alone to work less self-consciously and in his own time. The computer input the problems from prepared paper tapes and output instructions, questions, and information on the subjects' teletype. In return the subject was able to input questions and information through the teletype keyboard. The computer provided instantaneous and infallible responses to the subjects' questions.

The experimental session was controlled by a Macro 9 program (BLOCXT), which provided largely self-explanatory output text strings and required the subject to input only digits corresponding to the numbers of the blocks, 'space' and 'carriage return'. The keyboard of the teletype was masked so as to reveal only these keys (and 'rubout'). The hard copy output of the teletype was masked so as to reveal only the immediately preceding output. This would then disappear as soon as the teletype was used again.

It is difficult to be certain of the motivational effects, or lack of them, produced by the computer rather than the presence of the experimenter. Most subjects soon adjusted to the use of the teletype, and if anything seemed more relaxed than when directly observed by the experimenter. The BLOCXT program output the comment 'YOU ASKED A LOT OF QUESTIONS' if the subject asked seven or more questions. Without this facility it was thought that subjects might gradually increase their concept of a reasonable number of questions.

Each subject attended four one-hour sessions at the same time on consecutive days. The first session was used for training and in the each of the three remaining sessions subjects solved six problems with two examples of each type of rule (conjunction, affirmation, disjunction) in a randomized sequence. The experiment was conducted in two phases each using two matched groups of 18 subjects. In the first phase these groups were made up of fifth-year students of the Birmingham School of Architecture (A5) and science postgraduate students in their fifth year of study at the University of Aston (S5). In the second phase the groups were first-year architecture students in their first term of study (A1) and post A-level sixth-form school pupils eligible for university degree courses (6F).

4. RESULTS

4.1. First phase

The simplest measure of performance that can be made in the BLOCKS experiment is that of colour score. The subject has been asked to show as much of one of the colours (red/blue) as he can on the external vertical face of his solution. The vertical face is divided into fourteen modular bays; with the eight blocks coloured as in the experiment the maximum possible scores are 14 blue or 13 red, with no restricting structural rules. Each rule has its own maximum possible score varying between 11 and 14 for blue, and 10 and 13 for red.

The total differences between the maximum possible colour scores and those actually achieved by each subject (M–S) can be plotted against each of the three types of structure. The mean performance of the two groups of subjects, fifth year architects (A5) and scientists (S5) at each of the three types of problem, conjunction, affirmation, and disjuncton, is shown in Table 1. The results show the architects to do better with conjunctive rules than disjunctive rules. The scientists' performance is apparently quite different, being poor on conjunctive and good on disjunctive. It is interesting to note, however, that the grand mean score over all subjects and problems is almost identical for the two groups.

The difference between the subjects' performance at the conjunctive and disjunctive problems was examined separately for each group. In the case of the architects a two-tailed t-test for related samples showed a value of Student's t (df, 17) which

Table 1.
Mean colour score error, first phase subjects

| Problem type | Architects (A5) | | | Scientists (S5) | | |
| | Error type | | Total | Error type | | Total |
	PE	SE	(M–S)	PE	SE	(M–S)
Conjunction	0.89	1.17	2.06	1.56	1.39	2.95
Affirmation	0.44	1.33	1.77	0.61	1.00	1.61
Disjunction	0.50	2.44	2.94	0.50	1.33	1.83
Total over all problems	1.83	4.94	6.77	2.67	3.72	6.39

Grand total colour score error for all first-phase subjects 6.58.

has a chance probability of less than 0·05. The *t*-test just failed to show a significant difference between the two problem types for the science students. However, it must be recognized that the two results were in opposite directions.

A full analysis of variance was carried out on these results. Neither of the factors, problem type or subject group, were found to have a significant effect when examined separately. The interaction between them, however, was found highly significant (p<0·001). Inspection of the results shows nearly all this variation to be accounted for in the respective groups' handling of problems with conjunctive and disjunctive rules.

Why should subjects produce lower colour scores than the maximum, and why should these scores vary between subjects and problem types? It is possible to make two quite different kinds of error in the blocks experiment. A subject may fail to identify correctly the structural rule constraining him to use only certain combinations of blocks. However, to have produced an acceptable solution he must have unwittingly respected the constraint. In this case, as indicated by the results of the pilot study, the probability is that the subject was perceiving a more constraining rule, which is a subset of the actual rule. For example, a conjunction or affirmation may be incorrectly identified when the rule is in fact disjunctive. There is a second, more obvious reason for a subject achieving less than the maximum colour score. Having correctly identified the structural rule a subject may quite simply fail to arrange the blocks available to him in an optimal configuration.

In the pilot study subjects were asked at the end of each problem what they had thought the rule was. This practice was not continued in the main experiment for two reasons. Firstly it was thought that it attracted too much attention to the perception of structures and might tend to suggest a way of working that the subject might not otherwise have followed. Secondly the experimenter was not present in the laboratory and a more formal procedure for recording this information would almost certainly have affected the subject's attitude to the problems.

The on-line computer program which set the problems, ran and monitored the experiment was also capable of solving the problems. Where the subject had achieved a less than optimal colour score the program compared the subject's solution with the computer-generated optimal solution. The first stage of this analysis involved a search to discover if the optimal solution utilized the same four blocks as in the subject's solution. If this was the case then clearly all the subject need have done to achieve a maximal colour score was to rearrange the same four blocks. This type of error therefore is referred to here as a planning error (PE). In all other cases the less than maximal colour score is the result of one or more blocks being incorrectly chosen by the subjects. In these cases some block or blocks must be changed before the subject stands any chance of achieving the highest possible colour score. The use of one or more unfavourably coloured blocks indicates a lack of understanding by the subject of the actual structural rule in operation. This type of error therefore is referred to here as a structural error (SE).

It seems quite reasonable to hypothesize that the more spatially able architects would make relatively fewer planning errors (PE) than the scientists. This is confirmed by the direction of the results (see Table 1) but the difference proved to be just not statistically significant. A further experiment was therefore mounted to examine more closely the differences in planning ability between the architects (A5) and scientists (S5). Each subject was asked to plan all 16 possible combinations of the four pairs of blocks on the grid once to maximize red and once for blue. From these 32 solutions each subject was given a grand mean colour error score which depends entirely on planning errors since the structure is made explicit. The scientists showed a higher mean error score of 7·50 against the architects' score of 4·06, and this difference proved significant at the 0·001 level using a two tailed Mann-Whitney U test.

To return to the main experiment, it is noticeable that planning errors (PE) were higher for conjunctive problems than disjunctive problems. This is true for both subject groups although the scientists do comparatively even worse than the architects. It seems likely that this is due to the fact that conjunctive problems allow a smaller choice of blocks and therefore good planning becomes more critical. It may be that, in solving a disjunctive problem, a less than optimal solution can be improved either by exchanging one block for another or by replanning. In the conjunctive problem replanning may be the only option.

Turning our attention to structural errors (SE) reference to Table 1 shows that the architects made more structural errors overall than the scientists. This is particularly significant ($p < 0·01$, t-test) in the case of disjunctive problems. This is the complete reverse of the results obtained for planning errors. Here the scientists demonstrate their excellence at problem structure discovery. This ability is clearly likely to be more important in the case of disjunctive problems where it is recognised that the structure is particularly difficult to perceive (Bruner et al., 1956; Lawson, 1969).

4.2.   Second phase

The mean performances of the two second phase groups of subjects at the three types of problem; conjunction, affirmation, and disjunction, are shown in Table 2. Unlike the other subjects in the first phase of the experiment, neither group showed a preference for either conjunctive or disjunctive

Table 2.
Mean colour score error—second phase subjects.

| Problem type | Architects (A1) Error type | | Total | Sixth Form (6F) Error type | | Total |
	PE	SE	(M–S)	PE	SE	(M–S)
Conjunction	1·50	1·72	3·22	1·94	1·50	3·44
Affirmation	0·50	1·67	2·17	0·56	1·56	2·12
Disjunction	0·94	2·78	3·72	1·11	2·44	3·55
Total over all problems	2·94	6·17	9·11	3·61	5·50	9·11

Grand total colour score error for all second phase subjects 9·11.

problems. Both sixth form and first year architects showed an equally poor performance at both these problems types compared to the simpler affirmation.

It is noticeable that the second phase groups showed very significantly (*t*-test, p<0·001) higher overall colour score error scores than the first phase groups (grand mean 9·11 against 6·58). It can be seen that for each problem type the first year architects made fewer planning errors and more structural errors than the sixth form subjects, although none of these differences are significant.

4.3. Overall

Taken overall the results are most notable for the lack of structural errors (SE) made by fifth year scientists (S5) at disjunctive problems and the lack of planning errors (PE) made by fifth-year architects (A5) at conjunctive problems. A Kruskal-Wallis analysis of variance was carried out on the four groups for each of the six combinations of problem type and error type. Only the structural errors for disjunctive problems were found to be significantly different between groups (p<0·05). In the case of conjunctive problems a Mann-Whitney U-test showed that the fifth-year architects made significantly fewer planning errors than sixth-form students (p<0·02, two tailed).

4.4. Strategy analysis

The use of an on-line computer to control the experiment facilitates the analysis of subjects' strategies as well as final performance. Several statistics were computed to describe subjects' protocols, the two most interesting being reported here. The first statistic is the number of blocks in any question not found in the previous question. A grand mean number of blocks changed from the previous question was computed for each subject. (The first question of each problem was of course not counted.) These means tend to approach unity. It is most unlikely that a subject could score less than unity since this could only come about if over half of his questions were asked about exactly the same four blocks as their predecessors. This artificial lower limit makes the assumption of normality in the data unreasonable. The non parametric Mann-Whitney U-test was therefore used and showed that the fifth-year scientists (S5) changed significantly ($p<0.05$) more blocks between questions than the architects (A5).

The second statistic is the maximum colour score obtainable with the particular combination of blocks in each question. A grand possible colour score mean was computed for each subject over the first four questions of all problems. The variances of these means were also computed. These means are of course computed from different sample sizes for each subject so a non-parametric test had to be used for comparison between the groups. As with the blocks changed statistic the Mann-Whitney U-test was chosen. This revealed that for two-tailed tests the fifth-year architects' mean colour scores were significantly ($p<0.002$) higher and the associated variances significantly ($p<0.05$) lower than the scientists.

5.   DISCUSSION

These two results suggest that while the scientists were selecting the blocks in their questions in order to discover the structure of the problem, the architects were proceeding by generating a sequence of high scoring solutions until one proved acceptable. More detailed examination of the protocols reveals that most fifth-year science students did indeed operate what might be called a problem-focusing strategy, while most fifth-year architects by contrast adopted a solution focusing strategy. Neither strategy was so apparent amongst the second phase subjects, and none of the strategy statistics showed any significant difference between the two groups.

Interviews with S5 and A5 subjects conducted after the experimental sessions tended to confirm the existence of the two broad categories of strategy outlined here. However it must be pointed out that further experiments not reported here have suggested that both these general strategies have several variants and modifications. What was perhaps most surprising was the degree of rigidity with which most subjects said they adhered to their strategy. Indeed about half of each group could not conceive of any alternative to their methods.

It could be argued that the strategies of these two fifth-year university groups reflected the educational methods that they had undergone. An architect is taught mainly by example and practice. He is judged by the solutions he produces rather than the methods that he uses to arrive at them. Not so the scientist, who is taught by a succession of concepts and is only exercised by examples in order to demonstrate that he can apply these principles.

However, this is perhaps too simple an explanation. Both architect groups made fewer planning errors than the other two groups. Conversely the fifth-year scientists and sixth-form students tended to make fewer structural errors than the architects. Thus although there were not detectably significant strategy differences between the first-year architects and sixth-formers, the architects already showed greater ability in the production of solutions and less ability in the recognition of problem structure than the sixth-form sample. It could then be argued that the two sets of educational methods merely reinforced an already existent difference in approach between those who choose science and architecture careers.

There seems to be a very strong connection between strategy and performance. The fifth-year scientists gain over the other groups by making fewer errors at disjunctive problems. We have seen that this is due to a tendency to make fewer structural errors which would reasonably follow from a problem structure focusing strategy. The fifth year architects gain by making fewer errors at conjunctive problems due to a tendency to make fewer planning errors. This would reasonably follow from a solution planning orientated strategy. This seems to fit very nicely with previous findings and common sense. Concept attainment experiments show that disjunctive structure is more difficult to understand than conjunctive structure. In this experiment conjunctive problems must give rise to tighter planning situations than disjunctions due to the lower number of available alternative configurations. Thus both the scientists and architects show enhanced performance on just those problems in which one would expect their respective strategies to pay off.

6. SOME IMPLICATIONS

The approach to design suggested by the data presented here has also been found in other experiments not reported here (Lawson, 1972) and in the field by Darke (1978). Using research methods, derived more from sociology, based on interviews with renowned British architects, she has found a similar tendency to structure the problem by exploring aspects of possible solutions.

If the conclusions drawn from this empirical work are correct then this raises some difficult questions to be answered by the proponents of the view that design methodology can be understood as a special subset of scientific methodology. Such writers (e.g. Alexander, 1964) suggest that analysis and synthesis are discrete activities completed in that order. By contrast the conclusions here lend support to the more recent view of design put forward by Rittel and Webber (1973), who hold that such problems can never be completely described and are therefore not susceptible to total analysis. (This argument is well summarized by Cross, 1975.) If this is the case then it seems quite reasonable to suppose that designers would evolve a methodology which does not depend on the completion of problem analysis before synthesis can begin. This seems to be just what is observed in the approach taken by the fifth-year architects group in the experiment reported here. This raises important implications for the structure of computer-aided design systems (Lawson, 1971) and the drafting of legislative frameworks for designers such as the building regulations (Lawson, 1975).

Acknowledgments

The work reported in this paper forms part of a programme of research supported by grants from the Social Science Research Council and the Royal Institute of British Architects.

REFERENCES

Alexander, C. (1964), *Notes on the Synthesis of Form*, Harvard University Press, Cambridge, Mass.

Bruner, J.S., Goodnow, J.J. and Austin, G. A. (1950), *A Study of Thinking*, Wiley, New York.

Conant, M. B. and Trabasso, T. (1964), 'Conjunctive and disjunctive concept formation under equal information conditions', *Journal of Experimental Psychology*, **67**, 250–5.

Cross, N. (1975), *Design and Technology*, Open University, Milton Keynes.

Darke, J. (1978), 'The primary generator and the design process', in Ittelson, W. H., Albanese, C., and Rogers, W. R. (eds) *EDRA9 Proceedings*, University of Arizona.

Hunt, E. B. and Hovland, C. I. (1960), 'Order of consideration of different types of concepts', *Journal of Experimental Psychology*, **59**, 220–5.

Lawson, B. R. (1969), 'A study of set formation in design problem solving' (M.Sc. dissertation, University of Aston in Birmingham).

Lawson, B.R. (1971), 'Problem structure displays in computer-aided architectural design', *Ergonomics*, **14**, 519.

Lawson, B.R. (1972), 'Problem solving in architectural design', Ph.D. thesis, University of Aston in Birmingham.

Lawson, B. R. (1975), 'Upside down and back to front: architects and the building laws', *Royal Institute of British Architects Journal*, **4**, 25–8.

Lewis, B. N. (1963), 'Communications in problem solving groups', in Jones, J. C. and Thornley, D. (eds), *Conference on Design Method*, Pergamon, Oxford.

Neisser, U. and Weene, P. (1962), 'Hierarchies in concept attainment', *Journal of Experimental Psychology*, **64**, 640–5.

Rittel, H. W. J., and Webber, M. M. (1973), 'Dilemmas in a general theory of planning', *Policy Sciences*, **4**, 155–69.

3.4 The Psychological Study of Design

John C. Thomas and John M. Carroll

What is design? The most natural approach to defining design is to induce a definition from looking at examples of activities that are typically called design and fit the induced characteristics into a familiar framework. Since we are cognitive psychologists, the natural framework for design is problem-solving. In solving a problem, one starts with givens, performs operations and works towards a goal. Reitman (1965) distinguished what he called 'well-defined' problems in which the initial conditions, the operations, and the goal are well-specified from 'ill-defined' problems in which they are not. Theorem-proving in Euclidean geometry is paradigmatic of a well-defined problem. Designing a house, composing a symphony, or planning a research programme to study design seem to be obvious examples of ill-defined problems. Therefore, our first definition of design was as a particular class of problem-solving; one in which the goal, the initial conditions, and the allowable transformations are *all ill-defined*.

As we began and continued to work on the psychology of design, our notion of design changed from that of a TYPE of problem to a WAY OF LOOKING AT a problem. Any problem can be looked at as a design problem. For example, suppose someone sits down to prove a theorem in Euclidean geometry. Now, there are certain *human conventions* by which they may stay within that formal system. But, there is no law of the universe that says that they MUST stay within that formal system. If they VIEW the problem as allowing creativity, they may change the ground rules. They might decide that in order to prove the theorem, they will assume its converse and show

Originally published in *Design Studies*, **1** (1) (1979), 5–11. Reproduced by permission of Butterworth and Company (Publishers) Ltd.

how absurd the result would be. Such a person might end up with Riemanian geometry. The goal of the problem, in retrospect, was NOT well-specified. *A priori*, one would have thought the goal was to prove a particular theorem, but actually their goal was to find out something interesting.

Conversely, something which we typically think of as a design problem, such as designing a house, might be VIEWED otherwise. Suppose that an architect gives clients a questionnaire to fill out concerning their requirements for a house. Suppose further that this architect has a standard set of features and variations which are determined by the questionnaire results. In fact, this is NOT design. Much of what we call technological progress may be viewed as a process of rendering ill-structured design problems as more well-structured procedures for accomplishing the same ends—without requiring design.

We have been led therefore to a highly problem-solver-oriented problem-solving definition of design. For us design is a type of problem-solving in which the *problem-solver* views his/her problem or acts as though there is some ill-definedness in the goals, initial conditions, or allowable transformations.

MOTIVATIONS FOR STUDYING DESIGN

There are two distinct but related reasons that we are interested in studying design problem-solving. The first is to understand more fully how people think, and the second to discover how to help them design better.

Scientific: the extension of cognitive science

Psychology, at least experimental psychology, has tended to focus on simpler human functions. Thus, there is more work on sensation and perception than learning, more work on learning than on solving simple problems, and more work on solving simple and well-defined problems than on solving more complex or ill-defined problems. Our own bias is to focus more on complex human behaviour—on behaviours more particularly characteristic of humans.

As we designed and implemented our research on design, it became apparent that in most cases, effective design combines some top-down planning with bottom-up considerations. Experimental psychology (which can be conceived of as an attempt to design a theory of human behaviour) has implicitly concentrated almost wholly on studying humankind in a bottom-up fashion. First find out about simple things and then build those facts into a model of a human. Taken alone, this approach is almost certain to fail.

A complementary approach is to begin with an activity, like design, which encompasses an important way that humans actually seem to operate in the world. In trying to understand design, we can and do draw on what is known about how people remember, perceive, and think. But the motivating force is always attempting to understand something about a realistic, complex behaviour.

Applied: the need for better design

The world is changing faster now politically, economically, biologically, and technologically than ever before. As a consequence, we cannot simply allow our artefacts and procedures to evolve slowly (Jones, 1970). We will have to design solutions to problems, and we will have to do a better job faster.

Consider the important and difficult area of design which is software design. Software will probably constitute 90 per cent of the total data processing systems cost by 1985 (Boehm, 1973). Design is usually estimated to account for about one-third of the time and cost of producing software. In addition, however, testing and debugging account for about half of the man-hours involved in software development and between half and two-thirds of the errors originate during the design phase. Improvements in software design could be extremely important in lowering overall system cost. Software design is only one example of the increasing importance of design in our society; analogous cases could be made for the need for improved design in many other human endeavours.

METHODOLOGY FOR STUDYING DESIGN

Using a variety of approaches

We felt that a wide range of methods for studying design was appropriate, particularly at the beginning of our effort. After learning more about the relationship between methodology and what was likely to be discovered, a more intelligent choice of methodologies for finding out about particular aspects of the reality of design could be made.

Intuitive experience

Designers themselves have ideas about what is involved in the design process. These views are subjective—often ideological and sometimes idiosyncratic—while the scientific method strives for an objective and general view. Nevertheless, there should be a correspondence between the two. In order to understand this mapping, we feel it is necessary for us not only to try to understand scientifically (objectively) what designers do, but also to experience subjectively what designing is like. In addition to our experiences designing experiments, we are seeking a number of other design experiences so that we may see what designing is like from the 'inside' as well as from the 'outside' scientific view.

Scientific methods

With regard to the more scientific outlook, however, there are still a wide variety of possible methods for studying design. While the methods we have used vary in many dimensions, the most important dimensions can be collapsed into one of 'controlled/artificial' to 'uncontrolled/natural'. Towards one end of this continuum, we have videotaped situations where clients with real problems at IBM's T. J. Watson Research Center interacted with professional designers to attempt to solve that problem. Other than videotaping as non-intrusively as possible, we made no attempt to structure their interactions, their problems, their methods, or their solutions. The advantage of this method is that what we observe stands an excellent chance of being true of real design situations involving experts. The major disadvantages are that one cannot control the problems; cannot easily tell the extent to which the results are idiosyncratic; and the data analysis is subjective and difficult.

At the other end of this tradeoff, we have also conducted a number of experiments with a small-variance population (college students). We presented them with a relatively well-structured situation and systematically varied a number of factors. This allowed us greater experimental control and enabled us to use traditional statistical methods, but it is less sure how to generalize what was found to more realistic situations (Carroll *et al.*, 1978 a, b).

The methods and outcomes of these and other experiments are described in more detail below.

Expert experiments

It seemed wise before beginning more controlled experiments on design to observe what really went on in design. Then, if we decided to study some particular aspect, subprocess, or idealization of design, at least we would not confuse our method of studying design with real design.

Design dialogues

Our first attempt involved a standing request for professionals at IBM's T. J. Watson Research, when they had a problem that would require another expert to design something for them, to let us videotape their meeting. In one session, a head librarian was concerned about the implications of remodelling for access to computer terminals; in another, a department head wanted a better way of keeping track of names and addresses. A 'design dialogue' from a popular fiction work was also analysed and illustrated the same general format. These dialogues are dealt with more fully in Malhotra *et al.* (1978) and Carroll *et al.* (1978a).

Briefly, we characterized the dialogues as consisting of six states: goal statement; goal elaboration; solution outline; solution elaboration; solution testing (against the realities of the particular situation that had been originally ignored); and agreement or rejection of the solution. The word 'solution' can refer to the solution to the overall problem or to a sub-problem. The overall process can be viewed as a dialectic interaction between the information in the client's head and that in the designer's head. Typically, the client has two important kinds of information: a knowledge of his or her goals and a concrete knowledge of details of the particular situational context for the design-as-implemented. While the clients are more likely to be aware of what it is they know about the situation than to be completely aware of their goals, both kinds of knowledge are likely to be somewhat inexplicit. The designer, in contrast, knows less both about what is wanted and about the particulars of the situational context. The special knowledge that makes the designer an expert is about the relationships between partial goals and partial solutions. The states above evince an attempt by the client and designer to exchange enough knowledge to produce a solution that is feasible (can exist) and that is likely to help the client achieve the goals.

Software design

Two pilot studies have been done on software design. Based on our earlier analyses of client–designer dialogues, we determined that two important phases of the design process were: (1) the elaboration of a simple goal statement into more explicit and measurable 'functional requirements' and (2) from these, the elaboration of the general design solution.

In the first study (Malhotra *et al.*, 1978), four expert programmers were asked to write the functional requirements for a query system. We provided them with background data on the organization, the experience of the user population, the kinds of problems for which people would be using the system, and the systems environment within which the query system would operate.

The subjects spent an average of 3 hours each on the problem and gave four *very* different sets of functional requirements. One subject essentially explained why a standard query language should be used (Zloof, 1977). A second provided a menu of queries from which the user was to select. The third and fourth subjects both specified the functions and syntax of a query language. Although just a pilot experiment, the results are perhaps symptomatic of the extreme variability in software design.

In a second study (Miller, 1978) of software design, eight subjects were given a much more detailed statement of the functional requirements of the query language and asked to design the data structures and algorithms in eough detail for a PL/I coder. The designing took between 3 and 6 hours.

Solutions varied considerably in both content and form. The level of detail varied considerably between designers but also between different modules designed by a single person. In addition, various individuals took quite different approaches; viz., described different algorithms and different data structures. By definition, one would not expect identical solutions to design problems. What seemed particularly surprising about the result of the software design studies was the *fundamental nature* of the differences among experts both in translating general goals into functional requirements, and, given functional requirements, designing the algorithms and structures.

This suggests perhaps an overfocusing in software design on *quickly* getting *any* program that 'works' (runs) at the expense of other important but less quantifiable goals that would further constrain the solutions and thus result in more commonality in the solutions.

Controlled free response experiments

In this group of experiments, laboratory control of the task was established but subjects were relatively unconstrained in the form that their answer was to take.

Aids to problem solving studies

Part of our interest in design stems from a desire to produce methods of aiding people in their ability to solve problems. In the first of two experiments in which college students solved a series of problems, a structured aid was given to one group of college students. This aid required subjects to specify the goal of a problem, the objects, the transformations, and some of the relevant attributes. If anything, however, performance in the no-aid group was superior to performance in the experimental group.

The aid was based on our notion that people often fail to solve problems because they do not formulate the problems well. Although the aid failed to aid; ironically, it did give added support to our notion!! The answers that people gave to questions about goals, objects, and attributes showed that many people *did* slightly but vitally misunderstand the problems. For example, in one problem, subjects are to specify the minimum cards that need to be uncovered to *test* a statement about the cards (Wason and Johnson-Laird, 1972). Many subjects said that the goal of the problem was to *prove* the statement.

In the second study, we used an unstructured aid consisting of twenty pages of quasi-random words meant to stimulate ideas. Two design problems were included. On both, subjects given the word list produced significantly more creative solutions (Thomas *et al.*, 1977a).

Restaurant design

The design problems used in the problem aid experiments were so time-limited that it was arguable whether subjects were really designing. In an experiment which attempted to get at something closer to real design (though obviously still falling short), we gave a lavish amount of background data potentially relevant to designing a proposed restaurant which was to be in an old church. Each of the twenty-nine college students who served as subjects could communicate the design in whatever manner he or she chose. Half of the subjects were given the unstructured word list mentioned above. After completing their designs, subjects filled out a questionnaire which probed their goals, styles, and strategies.

One main purpose of this experiment was to explore and attempt to quantify the concept of creativity in design. Creativity has traditionally implied two notions: a creative design is not only unusual—it works. A creative design then implies both a practical design and an original design.

In order to define practicality, we specified the main goals of a restaurant, broke those high level goals down into subgoals, and kept refining until we reached a level we called 'functional requirements'. Each restaurant design was matched against each goal and a practicality score, P, was computed as the ratio of met goals to total goals.

A fairly objective originality measure, O, was also developed. Based on a feature analysis of the designs, the ratio of the information in each design to the maximum possible information was calculated and defined as originality. Thus, designs were considered more original to the extent they contained low-probability features.

Objective originality (O) was not correlated with the objective measure of practicality (P) overall. However, when subjects were divided according to SAT scores into six categories, there was a negative correlation between O and P within each group.

P and O were unrelated to any measures of the subject's strategy or style. There was, however, a significant relation for both to the sum of the subject's own ratings of the importance of various goals. Those subjects who expressed high concern with being original, imaginative, and novel scored low on P but high on O and conversely.

Those subjects who received the quasi-random word list scored significantly higher on P (but not differently on O) than those subjects who did not use the list. By looking at whether or not the subject was given this aid, at SAT score, and at expressed goals, a multiple correlation of 0.84 in predicting P was obtained (Thomas *et al.*, 1977).

In summary, objective measures of practicality and originality were defined on a theoretical basis. These were found to be related to the subject's retrospectively expressed goals. An unstructured word list was found to significantly increase practicality but not originality scores. The style of design and the subject's expressed strategy, on the other hand, were unrelated to either practicality or originality.

Letter-writing

As another example of a complex mental activity which can be viewed as a special case of design, we have looked at and analysed a number of letters. Two subjects were asked to think aloud while writing job application letters that were of very real importance to them. In addition, the first author looked at much of his own outgoing and incoming correspondence and classified letters as to purpose, and the extent to which the letter could have been produced by a fairly mechanical process. Letter-writing appears to be a sort of mini-design problem, at least in about half the cases. Apparently, a person seems to spend little time thinking about goals and planning out the approach before beginning. The designing seems to be going on during the actual writing process. Either the person is 'thinking ahead' at a strategic level while typing the words or, when the designing part becomes relatively difficult, the letter-writer may stop writing words altogether and concentrate on designing the strategy of the letter. These issues were explored in more detail in Thomas (1978a).

The main difference between letter-writing and other kinds that we have looked at is that the letter-composition seems to be much more bottom-up. New goals usually seem to be spawned by a consideration of why a certain word or phrase is not quite right. In contrast, in designing a restaurant, subjects seemed to have a fairly clear idea of what they were trying to accomplish before ever beginning to draw a floor plan.

Naming

The sorts of things we design in our ordinary use of language need not be as extensive and structured as letters. People routinely coin novel *ad hoc* name expressions for entities, events, and situations that lack established names: *a person you live with as if married but whom you are not married* ⇒ *a cohab*. We have recently begun to study naming by asking people to do just this—make up names for 168 novel referents composed of symbols, pictures, bits of text, steps in procedures, etc. (Carroll, 1979).

A number of very strong tendencies emerge. People seem to be far more compelled by 'descriptiveness' than by 'originality' in creating names. Very few true 'coined' forms occur (i.e. *cohab* and other such forms do not often occur). People tend instead to use existent lexical forms in novel combinations. These names are 'descriptive', conveying properties of the referent. Instead of *cohab* we might have *live-in lover*. We can interpret this preference for descriptive rather than coined names as a trade-off of practicality and originality, as in the restaurant study. In naming, our subjects favoured practicality, only rarely trading that for originality.

In subsequent studies, we have varied the goals of the name creation task. As in all design problems, the goals of naming are varied and they interact and trade off. In one study, ten subjects named a series of symbols with the instruction that names distinguish between referents while another ten named the symbols under the instructions that names are labels. Names designed in the two conditions differed completely: In the 'distinguish' condition, subjects designed names like *left one* for the leftmost of three identical symbols in a stimulus display. However, in the 'label' condition, the name generated was *boxed vee* (Carroll, 1979b).

The production of natural language is not often analysed as problem-solving, let alone as design, but we find this characterization to be quite natural and useful in describing the type of activity we have observed in both letter-writing and naming. Perhaps this view of language production will lead to theoretical advance in both the analysis of design and that of language production. The view that communication is, *in general*, best considered as design and interpretation (rather than transmission) is elaborated in Thomas (1978b).

Controlled experiments with restricted responses

Other research work with which we have been involved in the design area has made use of more circumscribed experimental model situations. In these studies we have maintained as many elements of design as possible, but restricted the response domain to facilitate scoring of success and characterization of strategies.

Isomorph studies

In one set of studies we were concerned with variables of problem statement and representation. Subjects were presented with one of two design problems that were isomorphic in structure: a schedule for stages in a manufacturing process or a layout for a business office. The latter version, the 'spatial' isomorph, involved designing an office layout to accommodate seven employees' offices within certain restrictions (some employees requested not to be adjacent to others, those with

higher status were to be placed near the end of the hall, etc.).
The 'temporal' isomorph involved scheduling the stages of a
widget manufacturing process such that certain stages precede
certain other stages, certain stages are simultaneous, etc. The
logical structure of both problems was identical, only the
statement of the functional design requirements was manipu-
lated.

In one experiment, subjects were given a matrix representation with which to record their design solutions. In another
experiment, they were not provided with any representational
aid—choice of representation was left entirely up to them.
With no representation scheme, subjects in the spatial isomorph condition were able to produce more adequate design
solutions in less time than subjects in the temporal isomorph
condition. When the representation scheme was provided, this
difference collapsed completely (Carroll *et al.*, 1978b).

In subsequent experiments we have tried to study not only
the final product of design problem solving of this type but also
the processes that determine that product. We refer to these as
studies of solution structure. Our analyses of solution structure
have for the most part also been conducted within the context
of the isomorph problem.

Library procedures study

Another scheduling problem we have studied involves organizing 22 tasks for maintaining a library among ten librarians. In a
sense this problem is like the client-designer dialogue situations
we described earlier, except that one role, that of the client, has
been simulated by a booklet. The subject receives discursive
definitions of the 22 procedures, and is told to treat the
librarians as if they were all identical.

However, while the problem is stated quite open-endedly, it
actually admits of relatively restricted responses. Nine of our
ten subjects assigned the tasks to the librarians by decomposing
the tasks into classes, and assigning each class of tasks to a
subset of the 10 librarians. We analyzed these classifications of
task finding high intersubject agreement as to which tasks
belonged together. Furthermore, the task classification seemed
to directly reflect underlying logical relations of similarity and
intuitive dimensions of separation between tasks. Finally, we
found evidence of design cycles quite analogous to those
identified in the client–designer dialogues. For additional
details see Carroll *et al.* (1978a).

OUTCOMES OF RESEARCH

In this section we will attempt to outline what we feel we have
learned from our study of design in terms of promising leads
and areas where more work is likely to prove fruitful.

Unstructured design aids

In two experiments, providing students with the unstructured problem-solving aid helped performance. The first author of this paper has used this aid on several real-life problems and also feels it is useful during an idea-production phase of solving problems. Further refinement and use of the aid, as well as further experiments to determine the conditions under which it is useful (Only for certain types of people? Only for certain types of problems? Only at certain phases of problem-solving?) would be valuable. Leonard (1968) reports the case of his daughter, whose drawing was stimulated by visiting a spectacular light show. Perhaps such a phenomenon is related to the hypnogogic imagery that often follows working with a particular kind of visual pattern that is attended to as in Robert Frost's poem *After Apple Picking*. It might prove interesting, then, to attempt to extend the aid to the visual domain and try 'flooding' people with possibly relevant visual ideas.

Decomposition of design problems

Our client–designer dialogue studies indicated that the overall attack on a design problem often organized into relatively smaller and simpler 'cycles': confrontations of portions of the total problem. An example from one of our dialogues appears below:

> *Client* … Now I've got some problems with where I place the printer, where the bloody control unit can go. I'd love to get the control unit buried under the floor somewhere.
> *Designer*: Uh-hu.
> *Client*: I don't know how minor that is. I think it's kind of minor.
> *Designer*: The control unit can be some 2000 feet from the scope so if you have an empty closet somewhere we can sort of hide it there so long as it is accessible.

Each cycle addresses a specific sub-problem or set of sub-problems constituent to the overall design problem (Alexander, 1964).

Sub-problems, aspects of the overall problem structure, and cycles, aspects of the solution structure, are related in complex ways. We certainly do not find that design cycles are of a standard duration or extent—that they are created by the participants with regard for problem structure. Nor do we find cycles that follow the strict hierarchic decomposition of the design problem, up and down the tree as it were, irrespective of duration and extent (these being what might be called functional as opposed to logical variables in the situation). *How* problem structure in part determines the cyclic structure of actual dialogues remains a provocative but wide open question.

Measures of creativity in design

We feel also that the work on measuring creativity via originality and practicality is promising. However, these notions need to be extended and refined. First, it should be pointed out that there are several quite distinct concepts of creativity. At the least, it is important to distinguish *manifest creativity* which a professional exhibits in real life from *creative potential* as measured by a test. Both of these notions are concerned with creativity as something which accrues to individuals.

Another view is that events or things exhibit creativity. This is similar to but distinct from process and product creativity. Our measures were only at best measures of the creativity of a *product*. In one study, we defined a 'prototypical chair' and then measured the creativity of a particular chair in terms of the number of features different from that prototype (Thomas *et al.*, 1977b).

In the 'restaurant' study we broke creativity down into two component measures: originality and practicality. There was evidence within the study that these measures had some validity and traded off against each other. This measure of originality also needs to be extended to consider higher order dependencies between features. In addition, what constitutes a feature can be defined in many ways. It may be worthwhile to do a 'sensitivity analysis' on future results to determine just how dependent any results on originality are on the particular definitions of what constitutes a feature.

Similarly, our measure of practicality, even in retrospect seems like a good one but should be further refined. In particular, in our study we let the relative importance of various goals and subgoals be chosen by the subjects. More reliable results on practicality may obtain if we were to ask each subject to weigh his or her goals before beginning to design and then see the extent to which each of these goals was met.

Designing in space versus time

Our work with design problem isomorphs begins to clarify the intuitive difference between temporal and spatial. We found a small but consistent conceptual difference (viz. spatial isomorphs are easier) but a notably strong tendency for spatial problem statements to encourage the use of graphic representational aids that were not encouraged by the temporal problem statement (Carroll *et al.*, 1978b).

This work suggests two levels of design aid: first, we could imagine reducing a problem in one domain to one in a conceptually simpler domain by systematic replacement of all terms and operators. This aid amounts to substituting problem isomorphs for actual target problems. It might even be feasible in relatively routine design areas. For example, one could imagine rendering a routine software design problem as some sort of spatial construction problem, presenting it to designers, and then retranslating the solution.

A more modest implication of this work has already found application in some design domains. We found that some problem statements implicitly suggest a representational aid better than do others. One might accordingly try to aid designers in these 'representationless' domains by merely providing them with ready-made schemes that they might otherwise not have considered. Flow-charting, in the temporal design domain of programming, is precisely an example of this—and one already widely implemented.

Salience of goals

As we have observed in several contexts, the typically implicit goal structure that the designer brings to the design situation will drastically alter his/her design activity and the product of that effort. In our study of naming we demonstrated the effect of goals by providing two differing instruction sets to our namers. When it was emphasized that names 'distinguish' objects in the world (Olson, 1970) subjects designed names that were indeed minimal distinguishers, like *one on the left*. When, however, it was emphasized that names 'label' objects in the world, subjects designed names that were not minimal distinguishers, like *boxed vee* as a name for one of three identical symbols.

This work does not tell us what naming is—distinguishing *or* labelling. We know antecedently that it can be both, and in context is probably neither. This work tells us that the particular goal structure brought to the design problem will provide a formidable and *systematic* influence on what naming becomes in that goal context. It makes the general point that the relative salience of differing goals will severely change design solutions. Of course, typically we do not know and cannot delimit the relevant goals in a design situation (Carroll, 1978).

Summary

To recapitulate briefly (and over-simply), the following findings seem both indicative of cognitive processes and suggestive of possible aids to the design process.

(1) An unstructured word list seems to help people generate ideas.
(2) Design PROBLEMS seem structured in terms of sub-problems.

(3) The SUB-PROBLEMS in design, however, are typically DYNAMICALLY produced during design; not completely specifiable at the beginning.

(4) The two main factors in a creative product—its originality and its practicality, are measurable and trade-off against each other.

(5) Designing in space seems easier than designing equivalent problems in time.

(6) A crucial aspect of design is specifying goals.

(7) Clients do not state all their goals explicitly; are probably not even aware of them before interacting with the designer.

(8) Goals stated in high level terms are not interpreted identically, even by experts in a field.

We have gained valuable insight from all the methods we have used so far. In the near future, we will probably continue to use all these methods but place greater emphasis upon controlled laboratory studies. In addition, however, we would like to expand our methods to include a longer-range, more complex field study and the development and testing of some aids to design in a real-world setting.

If anything, our original suspicion that there were similarities in design regardless of the domain has been strengthened after studying design with a variety of methods, tasks, and subjects. Activities as diverse as software design, architectural design, naming, and letterwriting appear to have *much* in common. Further communication between people studying design in diverse fields seems likely to be particularly fruitful.

Acknowledgment

This research was funded in part by the Engineering Psychology Programs, Office of Naval Research.

REFERENCES

Alexander, C. (1964), *Notes on the Synthesis of Form*, Harvard University Press, Cambridge, Mass.

Boehm, B. (1973), 'Software and its impact: a quantitative assessment', *Datamation*, p. 48.

Carroll, J. (1978), 'Names and naming: an interdisciplinary review', *IBM Research Report*, RC–7370.

Carroll, J. (1979a), 'Natural strategies in naming', *IBM Research Report*, RC–6533.

Carroll, J. (1979b), 'Towards a cognitive theory of reference', *IBM Research Report*, RC–7519.

Carroll, J., Thomas, J., and Malhotra, A. (1978a), 'A clinical–experimental analysis of design problem solving', *IBM Research Report*, RC–7081.

Carroll, J., Thomas, J., and Malhotra, A. (1978b), 'Presentation and representation in design problem solving', *IBM Research Report*, RC–6975.

Carroll, J., Thomas, J., and Miller, L., (1978), 'Aspects of solution structure in design problem solving', *IBM Research Report*, RC–7078.

Jones, J. C. (1970), *Design Methods*, Wiley, New York.

Leonard, G. (1968), *Education and Ecstasy*, Delacorte Press, New York.

Malhotra, A., Thomas, J., Carroll, J., and Miller, L. (1978), 'Cognitive processes in design', *IBM Research Report*, RC–7082.

Miller, L. (1978), 'Behavioural studies of the programming process', *IBM Research Report*, RC–7367.

Olson, D. (1970), 'Language and thought: aspects of a cognitive theory of semantics', *Psychol. Rev.*, **77**, 257.

Reitman, W. (1965), *Cognition and Thought*, Wiley, New York.

Thomas, J. (1978a), 'A cognitive model of letter writing', presented at Am. Psychol. Ass. Meet., Toronto.

Thomas, J. (1978b), 'A design–interpretation analysis of natural English', *Int. J. Man–Machine Studies*, **10**, 651.

Thomas, J., Lyon, D., and Miller, L. (1977a), 'Aids for problem solving', *IBM Research Report*, RC–6468.

Thomas, J., Lyon, D., Miller, L., and Carroll, J. (1977b), 'Structured and unstructured aids to problem solving', presented at Am. Psychol. Ass. Meet., San Francisco.

Thomas, J., Malhotra, A., and Carroll, J. (1977), 'An experimental investigation of the design process', *IBM Resarch Report*, RC–6702.

Wason, P., and Johnson-Laird, P. (1972), *Psychology of Reasoning: structure and content*, Harvard University Press, Cambridge, Mass.

Zloof, M. (1977), 'Query-by-example: a data-base language', *IBM Systems J.*, **16** (4), 324.

Part Four

The Philosophy of Design Method

Introduction

There is a school of thought within design methodology which believes that little knowledge of any real value will be gained simply from observing what designers do when they are designing. If one acknowledges that conventional design procedures are inadequate in some way, then what is needed is innovation of improved procedures rather than observation of faulty ones. New ideas for improved procedures are unlikely to arise merely from observation, it is argued. Instead, one needs as a starting point a tenable theory of design—a philosophy of design method—and it is this more theoretical view of designing which is addressed by the papers in this Part.

Hillier, Musgrove, and O'Sullivan's paper on 'Knowledge and Design' is concerned both with a philosophy of design method and with the related issue of the role of design research. Addressing architectural design specifically, they suggest that systematic design method was initially seen as a potential academic core for the discipline of architecture, with other research-orientated disciplines (e.g. building science) providing information to be used in a systematic design process. However, they consider that this has not in fact happened, and that an 'applicability gap' has opened between research and design.

In order to develop a new model of the role of research in design, and of the nature of design research itself, Hillier, Musgrove, and O'Sullivan refer to, and use as an analogy, developments in the philosophy of science. The conventional philosophical view-points of empiricism and rationalism both tried to eliminate the role of preconceptions in science, but now they argue, 'The question is not whether the world is prestructured, but how it is prestructured.' This new recognition in science is important for design, too: 'We cannot escape from the fact that designers must, and do, prestructure their problems in order to solve them.'

Their argument is that these prestructures—'the cognitive schemas which designers bring to bear on their tasks'—should be the primary concern of design research, and they draw on the philosophy of science as developed by Popper, Kuhn, and Lakatos in support of this argument. They suggest that the conventional view (the 'paradigm') of design research is based on two outdated notions: 'the notion that science can produce factual knowledge, which is superior to and independent of theory; and the notion of a logic of induction, by which theories may be derived logically from an analysis of facts'. It was these notions which had led to a view of a potentially rationalized design process which would analyse a problem into its component elements, add scientific information relevant to each element, and then synthesize a solution by some logical means. This inductive, analysis–synthesis view of design had taken hold, they suggest, because it embodied a traditionally liberal–rational sentiment that designs should be derived from an analysis of user requirements rather than from the designer's preconceptions.

Hillier, Musgrove, and O'Sullivan stress that science 'really operates' from prestructured view points and pursues investigations of the world from these viewpoints. Therefore it would not be unscientific for design to operate in a similar way. 'Why not accept', they ask, 'that only by prestructuring any problem, either explicitly or implicitly, can we make it tractable to rational analysis or empirical investigation?' They argue that 'design is *essentially* a matter of prestructuring problems' based on the designer's knowledge of solution types, knowledge of 'the latencies of the instrumental set'—i.e. the potential of the available technological means—and knowledge of 'informal codes' which relate users' needs to solution types and instrumental sets.

Like science according to Popper, design relies on conjectures, they suggest. Conjectures must necessarily come early in the design process, to enable the designer to structure an understanding of the problem and because a vast range of design decisions cannot be taken before a solution in principle is known. Conjectures become more sharply defined as relevant data are collected and used to test the conjecture. Conjecture and problem specification therefore proceed side-by-side rather than in sequence. Conjectures do not arise from data analysis, but from the designer's 'pre-existing cognitive capability—knowledge of the instrumental sets, solution types, and informal codes, and occasionally from right outside—an analogy perhaps, or a metaphor, or simply what is called inspiration'.

The model of a rational design process developed by Hillier, Musgrove, and O'Sullivan therefore has a core of conjecture–analysis, rather than analysis–synthesis, and a design develops according to this model by gradual refinement of an early conjecture. Among the advantages they claim for this model is that it avoids 'unworkable' notions of optimization from new data, it corresponds to the observed sequences and products of designing, it accepts that both data and conjectures are inherently incomplete, and it emphasizes the importance of the designer's prestructuring of the problem.

They go on to suggest that design research ('meta-design') should help designers to prestructure their problems; that is, it should concentrate on the 'codes' which link abstract functional requirements and instrumental sets. This would shift the emphases of design research away from study of procedures and towards the study of artefacts and their use. Its aim would be to give the designer a position of strength from which to make conjectures.

Hillier, Musgrove, and O'Sullivan's model of design is based substantially on Popper's 'conjectures and refutations' model of science. However, March, in his paper on 'The logic of design', comments that: 'The philosophy of Karl Popper has had some influence on modern architectural design theory. In the main its impact has been pernicious.'

The first part of March's paper (not reprinted here) is a criticism of Alexander's 'pattern language' approach to design. In particular, March challenges the view that 'the rightness or wrongness' of a design can be 'a question of fact, not a question of value', as claimed by Alexander and Poyner (see Chapter 2.2). Although equally concerned to develop a scientific approach to design, for March this would not be one which puts forward one 'correct' solution but one which 'selects a solution from a range of possibilities and seeks to assess its relative value'.

In referring to the influence of Popper, March stresses that 'To base design on inappropriate paradigms of logic and science is to make a bad mistake.' This is not a criticism of Popper's views, but a plea for design to be distinguished from logic and from empirical science. The distinctions are made thus by March: 'Logic has interests in abstract forms. Science investigates extant forms. Design initiates novel forms. A scientific hypothesis is not the same thing as a design hypothesis. A logical proposition is not to be mistaken for a design proposal.' According to March, design is in conflict with Popper's views on science, which oppose inductive logic, seek falsifiable hypotheses, and reject subjective probability statements. He says that Popper's criteria 'must be stood on their heads *in order to* maintain an approach which is rational' in design.

March introduces Peirce's concept of 'abduction' as a third mode of reasoning besides deduction and induction. According to Peirce, 'Deduction proves that something *must* be; induction shows that something *actually* is operative; abduction merely suggests that something *may be*.' March prefers to call abduction 'productive' reasoning, and specifies three tasks for rational designing: '(1) the creation of a novel composition, which is accomplished by productive reasoning; (2) the prediction of performance characteristics, which is accomplished by deduction; and (3) the accumulation of habitual notions and established values, an evolving typology, which is accomplished by induction.'

This leads to a 'PDI' model of a rational design process: an iterative procedure of production–deduction–induction. The first stage is the production of a design proposal. 'Such a speculative design cannot be determined logically, because the mode of reasoning involved is essentially abductive.' This proposal is based on an initial statement of requirements and on 'a presupposition, or protomodel'. The second stage is deductively to predict measures of the performance of the proposed design. The third stage is inductively to evaluate the proposed design and its predicted performance, and so to generate new or modified suppositions for an improved proposal.

Particularly in the early stages, there is some commonality between this model offered by March and that of Hillier, Musgrove, and O'Sullivan, despite their apparent differences over the relevance to design of Popper's views on science. The model of Hillier *et al.* consists essentially of prestructures–conjecture–analysis; whereas that of March consists essentially of presuppositions–conjecture–analysis–evaluation.

Analogies with, and distinctions between, design and science are also made by Broadbent in his paper on 'Design and theory building', who also draws on the works in the philosophy of science of Popper and Kuhn. He begins with Kuhn's concept of 'paradigms' in science, which he compares to the concept of 'style' in design. According to Kuhn, science undergoes a series of revolutionary changes in its paradigms. 'Normal' science is concerned with tackling the 'puzzles' within any paradigm; but occasionally a major re-interpretation—a new theory—gains converts and the paradigm shifts. 'Similar mechanisms are at work', Broadbent suggests, 'when "normal" designers are "converted" to the "theories" of a Gropius or Le Corbusier.'

Broadbent therefore sees design and designers in the same way as Kuhn sees science and scientists. He says that if scientists work within temporary paradigms 'then that must be the nature of human affairs', and so style and stylistic change are fundamental to design. But to avoid the problems which can arise from inappropriate styles (e.g. the environmental problems of the glass curtain–wall style in architecture) needs contributions from an adequate base of theory.

Broadbent goes on to discuss what a 'theory' is, and suggests two main criteria. Firstly, a theory is predictive—it offers a model which can predict the future states of the phenomena under investigation. Secondly, he draws on Popper's view that a theory is open to refutation. This concept allowed Popper to distinguish between what he regarded as genuine scientific theories and the social theories of Marx. Broadbent likens the 'theories' of Le Corbusier to those of Marx—such pseudo-theories have the power to change the phenomena with which they deal, whereas genuine theories do not affect the behaviour of their phenomena. 'The point obviously is that Marx and Le Corbusier changed the world because they wrote pseudo-theory and presented visions.'

Returning to design, Broadbent considers what might constitute theory in design. Some phenomena relevant to design can be the subjects of genuine theory—for example, materials' properties and perhaps human physiological responses to the environment. But he suggests that there cannot be any true theories of design as such, and so design will continue to be susceptible to pseudo-theories. However, these pseudo-theories must be subject to research which could expose their shortcomings.

Broadbent concludes by distinguishing between the designer and the scientist, and between designing and theory-building. He insists that the differences should not be regarded as a weakness of design, but perhaps quite the opposite: design activity is more difficult than scientific activity.

The particular nature of design knowledge is taken up by Daley in considering 'Design creativity and the understanding of objects.' In this paper for the 1982 conference on Design Policy, she tackles questions of design epistemology—i.e. 'the status of knowledge claims within the discipline'. What is the nature of design knowledge, and what is the nature of the metaknowledge of design theory? In particular, she asks whether design processes are wholly susceptible to systematic examination and to explication in conventional analytical terms.

Daley's starting point is that epistemology has a fundamental role to play in establishing an account of how it is that we understand our relationships with objects. 'Which is to say, how it is that we manipulate our conception of reality in such a way as to make innovations in spatial relations and, at times, create wholly new object configurations.' For Daley, our understanding of objects is inseparable from our knowledge claims about the design process. She therefore begins with an overview of classical epistemology, and the problem of the material existence of perceived objects. She reviews the rationalist and empiricist arguments, and Kant's view of perception as an active mental process. This leads her to seek to put any special cognitive abilities of designers to do with 'the imaginative manipulation of objects in space and time' within the general context of innate human abilities of perception and imagination.

One example of these innate abilities is that very young infants apparently 'assume' certain objects and shapes to exist, rather than others. Following Chomsky, there are also presumed to be innate generative principles underlying the ability to innovate in language. Daley develops a philosophical model—'an embryonic theory of mind'—to encompass these innate aspects of cognitive ability, drawing also upon Wittgenstein's concept of language as 'rule-following behaviour of a societal nature'. She proposes that 'the sense which we impose on our perceived experience, which centres on a world of continuous objects, and the value structures arising from sociality, are what makes knowledge possible'.

Daley goes on to claim that perhaps the propositional knowledge which is regarded as the principal content of intellectual activity represents only a small area of overlap of a set of knowledge systems. This leads her to some important conclusions about design, and about the limits to verbal discourse about design. In particular, she suggests:

> Only a relatively small (and perhaps insignificant) area of that system of knowing and conceiving which makes designing possible may be amenable to verbal description. To talk of propositional knowledge in this area, or to make knowledge claims about the thinking processes of designers, may be fundamentally wrong-headed. The way designers work may be inexplicable, not for some romantic or mystical reason, but simply because these processes lie outside the bounds of verbal discourse: they are literally indescribable in linguistic terms.

Where does such a conclusion leave design methodology? It seems to place some potentially severe—but as yet undefined—boundaries to any discourse about the way designers work, and it therefore suggests that many of the concerns of design methodology may be untenable. Perhaps a more positive way of viewing the outcomes of this whole set of philosophical papers is that they represent a freeing of design methodology from any naive adherence to science and to the ideology of science. Hillier, Musgrove, and O'Sullivan provide an updating on the epistemology of science and emphasize the necessity of cognitive 'prestructures'; March differentiates design from science and establishes the logic of abduction or productive thinking alongside induction and deduction; Broadbent identifies territories for design which make it more difficult and rewarding than science; and finally Daley claims types of non-propositional knowledge in design which are not amenable to verbal description. What are we left with, then, is a greater confidence in design as a way of knowing and thinking, a new set of axioms for design methodology, and a recognition that much of the discourse must be transferred to new and more fundamental levels of epistemology.

FURTHER READING

A major reference in the debate on the relationship between science and design is Simon, H.A. (1969), *The Sciences of the Artificial*, MIT Press, Cambridge, Mass. (second edition, 1981). Two of the editorial chapters in the proceedings of the 1965 Birmingham Conference are particularly relevant: Gregory, S. A. (1966), 'Design and the design method' (Ch. 1) and 'Design science' (Ch. 35), in Gregory, S. A. (ed.), *The Design Method*, Butterworths, London. Many papers are related to this topic in the 1980 Portsmouth conference proceedings: Jacques, R., and Powell, J. (eds) (1981), *Design: Science: Method*, Westbury House, Guildford, including Broadbent, G., 'The morality of designing', and Cross, N., Naughton, J., and Walker, D., 'Design method and scientific method' (also published in *Design Studies*, 2, 4, pp. 195–201). See also Batty, M. (1980), 'Limits to prediction in science and design science', *Design Studies*, 1 (3), 153–9.

A philosophical critique of design methods in architecture is provided in Chapter 2 of Scruton, R. (1979), *The Aesthetics of Architecture*, Methuen, London.

4.1 Knowledge and Design

Bill Hillier, John Musgrove, and Pat O'Sullivan

Research of one kind or another has now a longish history in building. By and large, this increased investment in research has proceeded side by side with a marked deterioration in the quality of building. A serious 'applicability gap' appears to exist. Regardless of the quality of research work itself, the history of attempts to link research to improvements in environmental action is largely one of confusion and failure (RIBA, 1970).

Ten years ago, when the ground was being cleared for a great expansion of architectural research activity, programmatic statements took a clear line. Design was a problem-solving activity, involving quantifiable and non-quantifiable factors. Research, it was thought, should bring as many factors as possible within the domain of the quantifiable, and progressively replace intuition and rules of thumb with knowledge and methods of measurement. This process would never be complete. Non-quantifiable elements would remain. In order to assimilate such knowledge and use such tools as we were able to bring to bear on design, the procedures of designers would have to be made more systematic. Because the education of architects was broad and shallow, and because they were concerned with action rather than knowledge, they could not be expected to generate new knowledge for themselves. This was the job of 'related' disciplines, whose concern was the advancement of knowledge. Architects, on the other hand, knew about design, and should make systematic design their research focus. Otherwise their contribution to research lay in

Originally published in Mitchell, W. J. (ed.) (1972), *Environmental Design: Research and Practice*, University of California, Los Angeles. Reproduced by permission of the Regents of the University of California, Los Angeles. © 1972 by the Regents of the University of California from whom permission to reproduce any part of this article should be sought.

technological development, or as members of multi-disciplinary teams, in defining the problems for others to solve.

The educational consequences of these notions were that schools of architecture and planning were to be located in an education milieu containing a rich variety of related disciplines, and students were to be well grounded in each of them. The core of the architectural course would still be design, and, at the academic level, this meant increased concentration upon systematic methods. Students would be taught to analyse problems, and to synthesize solutions.

A few voices crying in the wilderness that architecture contained its own fundamental disciplines could not stop the onward march of these simple and powerful ideas, and by and large, they still hold the stage today. But if these are to be paradigmatic ideas by which we define our subjects and link them to action, then today's landscape (although promising in that other disciplines are developing their latent 'environmental' interests), must appear depressing. Systematic design studies are in disarray. Increasing numbers of research workers, including architects, are moving into the areas previously called 'unquantifiable'. There is a widespread feeling that an 'applicability gap' has developed between research and design. Design is still led by the nose by technology, economics, and imagistic fashions. The human sciences and architecture are still at loggerheads. Education, with few exceptions, has not managed to develop a radically new capability in the problem-solving power which students bring to design.

In fact, we are far from pessimistic about the progress of architectural research, largely because a great deal is now happening that cannot be explained in terms of the ideas we have outlined above. The situation has outstripped the paradigm which gave birth to it. But the intelligibility of the situation is poor, perhaps because it is inconsistent with the paradigm. We require some radical overhaul of its assumptions—particularly those to do with the relationships between knowledge and design, and the presupposed polarities (e.g. rationalism/intuitionism) along with a new effort to externalise the dynamics of the new situation. To us this seems to be an essential step before the 'applicability gap' is compounded by a 'credibility gap' arising from the gulf between what is expected of research, and what research appears to be offering.

Perhaps the simplest way of introducing what we have to say is by drawing an analogy with the slow but decisive shift in philosophy and scientific epistemology over the past half-century or so. Implicit in both the rationalist and empiricist lines of thought was the notion that in order to get at truth, preconceptions must be eliminated or at least reduced to the minimum. Rationalism began its long history by proposing *a priori* axioms whose truth was supposed to be self evident; empiricism relied on the neutrality of observation. Since the early part of this century, developments in such areas as psychology, metamathematics, logic, and the philosophy of science have combined to show that both of these are impossible and unnecessary to an account of scientific progress. Far from being removed from the field of science, the cognitive schemes by which we interpret the world and prestructure our observations are increasingly seen to be the essential subject matter of science. The question is not whether the world is prestructured, but how it is prestructured.

Too often these developments appear to have escaped the attention of scientists working in the environmental field and designers interested in research and looking to research for solutions to problems. This is particularly unfortunate, because the idea of prestructuring has immediate and fundamental applications in design. We cannot escape from the fact that designers must, and do, prestructure their problems in order to solve them, although it appears to have been an article of faith among writers on design method (with a few exceptions; Colquhoun, 1967) that this was undesirable because unscientific. The nub of our argument is that research in the field of the built environment and its action systems should see as its eventual outcome and point of aim the restructuring of the cognitive schemes which designers bring to bear on their tasks, not in terms of supplying 'knowledge' as packaged information to fit into rationalized design procedures, but in terms of redefining what those tasks are like, and using the heuristic capability of scientific procedures to explore the possible through a study of the actual. It is our view that the notion that well-packaged knowledge coupled with a logic of design can lead to radically better artefacts, on the evidence we have, should be relegated to the realm of mythology. But in arguing our case we would like to say a little more about why we think modern scientific epistemology has an important bearing on design and meta-design (which we will argue is probably the simplest and most adequate characterization of design research) and why it can help us reconstitute our paradigmatic notions about the subject.

Fifty years ago, it was still possible to think of science as ultimately constituting a set of signs which at the most rudimentary level, would bear a one to one correspondence with atomic facts, and that these could eventually be combined by the laws of induction and verification into a pyramid of laws of greater and greater generality. Scientists on the whole believed this to be the case, and philosophers concerned themselves to show how it would be accomplished.

The overthrow of the Newtonian account of the universe, previously taken as the paradigm of positive knowledge arrived at by observation and induction (as described by Newton himself) threw scientific epistemology into a crisis, the effects of which are still with us. We will give a brief account of this later on. Shortly afterwards, even more remarkable and undermining developments took place in the foundations of mathematics and logic (Keene, 1959). Gödel (Nagel and Newman, 1959) showed, by his incompleteness theorem, that 'the construction of a demonstrably consistent relatively rich theory requires not simply an "analysis" of its "presuppositions", but the construction of the next higher theory with the effect', to continue quoting Piaget, that:

Previously it had been possible to view theories as layers of a pyramid, each resting on the one below, the theory at ground level being the most secure because constituted by the simplest means, and the whole poised on a self sufficient base. Now however 'simplicity' becomes a sign of weakness and the fastening of any storey on the edifice of human knowledge calls for the construction of the next higher theory. To revert to our earlier image, the pyramid of knowledge no longer rests on its foundations but hangs by its vertex, an ideal point never reached and, more curious, constantly rising. (Piaget, 1971)

This is of vital importance, not simply because it demonstrates the inherent limitations of formalism, and the impossibility of such notions as the class of all classes, or the single unified science, but because it demonstrates that there is a necessary hierarchy which limits what we can mean by knowledge—the hierarchy of meta-theories and meta-languages, independent of (we can think of it as orthogonal to) the hierarchy of levels of integration of phenomena in the 'real' world, which constitutes the formal basis of most scientific disciplines. Any cognitive formalization takes a lower-order formalism for its object and can itself become the object of a higher formalism. To quote Piaget (1971) again:

The limits of formalism can, more simply, be understood as due to the fact that there *is* no 'form as such' or 'content as such', that each element—from sensory motor acts through operations to theories—is always simultaneously form to the content it subsumes and content for some higher form.

If we accept that the idea of a monumental edifice of knowledge, descriptive of the world in its account of facts and explanatory of them in terms of theories of increasing generality, has to be given up, what have we left? Have we not effectively debunked the idea of knowledge? Having got rid of positivism, are we left with pure relativism? Intuitively, we feel that such a retreat cannot account for the success of science in improving our understanding of the world and our capacity to modify it. If we adopt a position of pure philosophic relativism then relativity (in theoretical physics—we are short of terms here) and the atomic bomb appear as a kind of epistemological paradox. If on the other hand, we accept that there are strong reasons for rejecting both positivism and pure philosophic relativism, then where do we go? It seems that, as with Scylla and Charybdis, we cannot escape the one without falling into the other.

It is against this background that the achievement of scientific philosophers like Karl Popper, Thomas Kuhn, and Imre Lakatos take on their full stature. Popper (1959, 1963) has demonstrated that a logic of induction and the principle of verification, previously the twin pillars of positivist science, were both unattainable and unnecessary, and that science could be contained within a hypothetico-deductive scheme; Kuhn (1962) suggests a changing epistemological paradigm, within which science can operate as a puzzle-solving activity until the next revolutionary 'paradigm switch'; Lakatos (Lakatos and Musgrave, 1970) reconstructs science as conflicting sets of interrelated theories (on a smaller scale and more volatile than Kuhn's paradigms), retaining the idea of a 'negative heuristic theoretical core' and a 'positive heuristic' puzzle-solving area, each of which exhibits at any time either a 'progressing' or 'degenerating' problem shift according to whether or not it is able to predict new phenomena within its basic theories without having to add *ad hoc* hypotheses to account for newly discovered phenomena. Then we have a reconstruction of science which is able, in a highly non-linear way, to account for its own continuity, as well as offering some rational justification for using the word 'knowledge' perhaps, to use Popper's expression, as 'piles in the swamp', the swamp being essentially the infinite regress of meta-theories and meta-languages.

The simplest reconciliation of these lines of thought in meta-mathematics and the philosophy of science is to state frankly that the object of science is cognition, and that it is the strategems of science that are directed towards the real or empirical world. More precisely, we could say that science is about 'remaking cognition', it being clear that if we were satisfied with our cognitive codes for deciphering the world, we would not have science. This seems to us an adequate resolution of the old philosophical problem of whether the 'world out there' or our perception of it is the more real. Such a definition is implicit in the work of psychologists like Kelly (1964), who characterize everyday behaviour by analogy with scientific behaviour. It is a small step to reverse the argument, and it allows us to account not only for the pre-occupation of science with the empirical, but also for the fact that some advanced areas of science—notably certain branches of theoretical physics— have had no means of contacting the empirical world for about forty years. We would hardly be satisfied with a characterization of science which relegated theoretical physics to the realm of metaphysics.

How does all this help us with architectural research? First, it should be clear that once we move away from the establishment of basic criteria set up with a view to avoiding physical discomfort (which we knew how to do anyway in pre-scientific days) then we can avoid a lot of misconception about the status of 'knowledge' in design. Secondly, we can begin to see the problems raised by the paradigm for research in architecture that we outlined at the beginning of the paper. Thirdly, it provides us with a better method of making fertile analogies between and thus in connecting the activities of scientists and designers.

The paradigm we suggest as underlying most current research activity in architecture appears to be based on two notions about science that take no account of the developments we have outlined: the notion that science can produce factual knowledge, which is superior to and independent of theory; and the notion of a logic of induction, by which theories may be derived logically from an analysis of facts. In the paradigm, these two notions appear to constitute the fundamental assumptions on which the whole set of ideas is founded: first, that the role of scientific work is to provide factual information that can be assimilated into design; second that a rationalised design process, able to assimilate such information, would characteristically and necessarily proceed by decomposing a problem into its elements, adding an information content to each element drawn as far as possible from scientific work, and 'synthesizing' (i.e. inducting) a solution by means of a set of logical or procedural rules.

So far we have suggested very theoretical reasons why such ideas would not be viable or realisable. But equally, from the more practical point of view of the designer or the student, the ideas—or more precisely the operational consequences that flow directly from them—appear even more unviable. Designers are left to make their own links with research by assimilating 'results' and quantification rules, and to evaluate them as they appear without guidance on priorities or patterns of application. The designer's field thus becomes *more* complex and *less* structured. It follows that if a designer cannot make use of this 'information' he is forced to the conclusion that it is because his procedures are not systematic enough, with the result that if he tries to improve himself, he immediately becomes preoccupied with means at the expense of ends.

Similar consequences flowed from these twin paradigmatic assumptions in architectural research itself. For example, building science as a university discipline tended to remain separate and independent of the design disciplines, usually as a research-oriented service-teaching department, sometimes even generating the packages of knowledge that were to fit into the rationalized design procedures. In trying to formalise the process, designers were forced into developing concepts like 'fit' and 'optimization' simply in order to complete the line of logic by which a 'synthesis' could be accomplished, even though such notions are highly artificial in terms of what buildings are really like and are actually refuted by considering buildings as time-dependent systems rather than as once-and-for-all products.

Our negative aim in this paper was to try to show why the advance of research related to design has so far appeared to progress in parallel with deterioration in the acceptability of the designed product—and this, in the UK, in spite of two decades of excellent work by such bodies as the Building Research Station and government departments, well disseminated in intelligible form and often containing mandatory requirements. We hope that we have shown that there are both theoretical and practical reasons why such a state of affairs should not surprise us. If the present paradigm is unworkable in its essentials, what can we put in its place? We have to preface our proposals with some suggestions about the nature (the actual nature as well as the desirable nature) of design activity.

It is not hard to see why the analysis–synthesis, or inductive, notion of design was popular with theorizers and even with designers as a rationalization of their own activities. The architectural version of the liberal–rational tradition was that designs should be derived from an analysis of the requirements of the users, rather than from the designer's preconceptions. It is directly analogous to the popularity of induction with scientists who were anxious to distinguish their theories as being derived from a meticulous examination of the facts in the real world. The point we are making in both cases is not that the ideas are immoral or fundamentally deceptive—scientists *do* describe meticulously the 'facts' of the situation, and designers *do* pay attention to the details of user needs—it is that they are theoretically untenable and unnecessary, and as a result, practically confusing.

The first point we would like to make about our version of science in relation to design, is that if scientists really operate by a kind of dialectic between their prestructuring of the world and the world as it shows itself to be when examined in these terms, then why should such a procedure be thought unscientific in design? Why not accept that only by prestructuring any problem, either explicitly or implicitly, can we make it tractable to rational analysis or empirical investigation?

The second point is also in the form of a question. If rationality in design is not to be characterized in terms of a procedure that allows the information to generate the solution, then in what terms can it be characterized? Is it a redundant notion? Is there any alternative to the mixture of intuitive, imitative, and quasi-scientific procedures which appear to characterize design as it is carried out? We would like to work towards answers to both of these questions by using some of the ideas we have discussed in a kind of thought experiment about the nature of design.

First, some observations about reflexivity (cognitive activity making itself its own object, or part of its object) and meta-languages and meta-theories (cognitive activity making other cognitive activity its object). These, it would appear, have clear parallels at the social level, in terms of the progressive differentiation of roles, especially in areas like design where physical activity is preceded by cognitive and reflective activities. For example, if we start with a simple picture of a man making an object, then it would be reasonable to argue that in as much as he has a definite cognitive anticipation of the probable object (i.e. he is not simply experimenting by trial and error with the latencies of his tools and raw materials) then he is acting analogously to a designer as well as being a maker. His cognitive anticipation of the object is part of the field of tools and raw materials that constitute his 'instrumental set'. Design as we know it can be seen as the socially differentiated transformation of the reflexive cognition of the maker in terms of the latent possibilities of his tools, materials, and object types. Its object is not the building, but at one remove, sets of instructions for building. The activity called architectural research can be derived by an exactly similar transformation, namely a socially differentiated transformation of the reflexiveness of the activity of design upon itself i.e. its object is design, and its product takes the form of rules or rule-like systems for design which stand in the same relation to design as design does to building. As in other sciences, it finds the best way of doing this is by addressing most of its strategies to the 'real' world, and if we are not careful this coupled to the fact that the activity is necessarily multi-disciplinary, tends to conceal the 'deep structure' of the activity. This is why we suggested earlier that we should call the research activity meta-design. At least this might begin to emancipate us from the silly (but pervasive) idea that the outcome of research is 'knowledge', to be contrasted with the absence of such 'knowledge' in design.

We can perhaps clarify the characteristics of design as a cognitive activity by going back to the very simplified situation we have just referred to, to see if we can discover what there is in the maker/designer's field, and go from there to see how it differs today. Here we owe some debt to Levi-Strauss' (1966) discussion of 'bricolage' as an analogy to myth making.

We can imagine a man and an object he will create as though separated by a space which is filled, on the one hand, with tools and raw materials which we can call his 'instrumental set' (or perhaps technological means), and on the other a productive sequence or process by which an object may be realized. If time is excluded from the space, we can conceive of the 'instrumental set' as though laid out on a table, and constituting a field of latencies and preconstraints. If time is in the space, then the instrumental set is, as it were, arranged in a procedure or process.

The total field thus exhibits two types of complexity, and we may allow that the maker is capable of reflexively making both types of complexity (the latencies of the instrumental set, and the distribution in process-time) the objects of his attention.

Two basic strategies appear to be open to him. He can either redistribute the latencies of the instrumental set in process-time according to some definite cognitive anticipation of the object he is creating, i.e. pursue a definite design or plan, which may be based on an analogy or on pure imagination, as it may be conceived in terms of the familiar products of the instrumental set. Or he can, as it were, interrogate his instrumental set, by an understanding of its latencies in relation to general object types. In both strategies an understanding of the latencies of instrumental sets and a general knowledge of solution types is of fundamental importance. In other words, the maker's capability in prestructuring the problem is the very basis of his skill, even if he wishes to proceed heuristically by interrogating his instrumental set and exploring unknown possibilities by a dialectic between his understanding of the latencies and limitations of the instrumental set and his knowledge of solution types. On this basis we would argue that design is *essentially* a matter of prestructuring problems either by a knowledge of solution types or by a knowledge of the latencies of the instrumental set in relation to solution types, and that this is why the process of design is resistent to the inductive–empiricist rationality so common in the field. A complete account of the designer's operations during design, would still not tell us where the solution came from.

But there is an escape clause. As with science, it is not a matter of *whether* the problem is prestructured but how it is prestructured, and whether the designer is prepared to make this prestructuring the object of his critical attention. From here we would go on to suggest that the polarization we have assumed between rational and intuitive design should be reformulated as a polarity between reflexive design (i.e. design which criticizes its understanding of the latencies of instrumental sets and solution types) and non-reflexive design (i.e. design which is simply oriented towards a problem and which therefore operates within the known constraints and limits of instrumental sets and solution types). To equate rationality with a certain type of systematic procedure appears therefore, quite simply, as a mistake.

It is obvious that today the designer operates in a field which is considerably more complicated than the one we have described, based on a man making an object. The notion of prestructuring is *necessary* to any conceptualization of design, but not *sufficient* in itself. We have to look at the complications and how they have evolved, in order to complete our conceptualization of the designer's field and his operations in it.

The most obvious difference is that design is not simply the reflexive/cognitive aspect of making an object, but a separate, socially differentiated activity with its own internal dynamic and its own end product, namely sets of rules for making artefacts. It is also a highly specialized activity, carried out by a clearly defined social group. There is therefore no direct link between interrogating the instrumental set and the result as it is likely to be experienced by those who use it. We thus require a great deal of information about the latter in order to interrogate the instrumental set.

We can explore the consequences of this development by trying to imagine what life was like when we had designers, but not user-requirement studies. How did we live without them? The answer seems quite simple. Notions about the user were built into the instrumental set and the solution types. The instrumental set was comparatively unsophisticated and had in any case been developed mutatively over a long period. It was already an expression of the basic physiological requirements of users in terms of available technology, and probably a reasonable approximation of their psychological and other expectations. The solution types had been similarly evolved, and contained already the notions of use and activities within the building. We could say that, contained in the instrumental set and the solution types was an implicit, historically evolved code, which linked the means to the ends. It would be difficult to decipher and reconstruct, but we can see that it was there, and, in principle, how it got there.

Since those days we have seen developments like the proliferation of building types, and the proliferation of instrumental sets (technological means) and a formal organization of the process which results in most activity being of a one-off kind with the simple effect that the users' needs in terms of activities, physiological requirements, and cultural expectations are no longer contained, as it were, in the instrumental sets and solution types. A much freer, more indeterminate situation appears to exist. This deficiency is made up in terms of information which is expressed in terms of the users rather than in terms of buildings, and the designer operates a kind of *informal code* for linking one to the other. Part of the outcome of research in the past has been a piecemeal and atomistic partial replacement of the codes, by formal rules which when implemented often have the unfortunate effect of dictating the whole design (the 2 per cent daylight factor is a classic example). The designer's task becomes something like the utilization of these codes in order to link the information he gathers about the project to his interrogation of the increasingly prolific instrumental sets, or his manipulation of solution types. He has to deal similarly with the proliferation of information extraneous to the particular problem relating to standards, constraints, quantification rules, etc. In this situation it is perhaps no wonder that the designer (unless his ambitions are frankly artistic) welcomes the prospect of a logic whereby solutions can be synthesized out of information. It offers him the prospect of eventual escape from the contradiction of actually working by the interrogation of instrumental sets or the adaptation of solution types, as he always did, but being expected to utilize a procedure of optimizing information which bears little relation to building, except where piecemeal atomistic rules have been developed. Perhaps we should add one more point to this analysis: that the informal codes the designer must use to link information to built outcomes are also instances of problem prestructuring.

If this is a reasonable characterization of the principal elements in the designer's field, then at least we are some way to understanding why designers do not produce better buildings out of the information research provides, and why, with expanding technological means and user requirements, the theoretical open endedness of architectural problems lead to so little fundamental variety in the solutions proposed. With a proliferation of poorly understood instrumental sets, increasingly masked by unrelated information, we would expect that a retreat to the most basic form of prestructuring—the adaptation of previous solutions—would become the only viable way through the morass. Far from helping the designer escape from his preconception, the effect of proliferating technology and information is to force the designer into a greater dependence on them. Innovation becomes more rather than less difficult, but the diffusion of uncritical innovation would become more rapid. A situation develops in which a few experiment and others adapt solution types, without understanding or evaluating the rationale of the original experiment. The net result is unstructured innovation, with slow and piecemeal feedback, giving the impression of arbitrary shifts in fashion. This seems a not unreasonable account of the situation we have, and would explain why even well disseminated and well presented information—such as widely exists in the UK from BRS and government departments—either does not lead to an improvement in the product or does so only in a haphazard way.

We would also suggest that this leads to a situation in which students are learning two different and largely unrelated strategies: methods of analysing a problem into its elements; and a knowledge of informal codes and solution typologies, which they pick up almost as by-products of architectural education, and which act as the prestructuring that enables them actually to design buildings.

We have argued that the chief elements present in the designer's field are *knowledge of instrumental sets, knowledge of solution types, informal codes, and information.* These cannot usefully be reduced to homogenized 'information', although it is possible at a theoretical and formalized level. Now we would like to use these ideas to try to construct a lifelike conceptualization of design as an activity.

These elements constitute the designer's field, his set of latencies and preconstraints. Somehow these are to be distributed in a process-time. We will need to introduce one or two further basic ideas as we proceed, but we hope that these will either be from those we have already discussed, or simple logical statements of an unproblematical kind.

For example, it seems unproblematic to say that when a design problem is stated there are, theoretically at least, a number of solutions open, probably a very large number. Yet only one of these possible solutions will be the final one that is built. We may reasonably say that some process of *variety reduction* has taken place. The variety of possible solutions has been reduced to one unique solution by some means. The succession of documents produced during design reflect this progressive reduction of variety. More and more specific drawings, for example, exclude more and more detailed design possibilities. We would like to introduce this as a basic idea in our conceptualization of design.

A second idea we would like to introduce is that of *conjecture*. Here we would like to go back to science. It was once thought that conjecture would have no place in a rigorous scientific method. It was thought to be akin to speculation, and science sought to define itself in contradistinction to such notions. Since Popper we know that science cannot progress without conjecture, in fact that together with rigorous means of testing, conjectures constitute the life blood of science. Conjectures come from anywhere, and because they are not derived from the data by induction, it does not mean that the process of thought of which they form part is any the less rational or rigorous. What is irrational is to exclude conjecture. So we will include it in design.

How does the reduction of variety from many possible to one actual solution take place? Obviously anything we can say here will only be an approximation of any particular case. But our aim is to try to understand the process of design as it exists in the real world, in order to try to define the contribution of meta-design. What we are aiming at is some more or less true to life approximation of the psychology of design, bearing in mind that design is a practical as well as a cognitive activity, and that design problems do not happen in a social vacuum, but are socially constructed.

Beginning with a theoretically open problem, with an unlimited number of solutions, it should be clear that the variety of possible solutions is already reduced before any conscious act of designing begins by two sets of limiting factors, one set external to the designer, the other internal. The first set we can call 'external variety-reducing constraints' and these can often be quite powerful, or even totally deterministic of the design. For example a client who says categorically, 'I want one like that' has already reduced the number of possible solutions to one. More often the external constraints will be of a less overt, but still powerful, kind, such as norms of appearance, availability of technological means, costs, standards and so on. Some of these will not be fully understood by the designer at the outset, but as he specifies them their role as variety reducers will become clearer.

The second set we can call the 'internal variety reducers' and these are an expression of the designer's cognitive map, in particular his understanding of instrumental sets and solution types. This notion of the pre-existing cognitive map is very important indeed, because it is largely through the existence of such maps that any cognitive problem-solving activity can take place. They are, and must be, used by the problem-solver in order to structure the problem in terms in which he can solve it. It acts as a kind of plan for finding a route through problem material that would otherwise appear undifferentiated and amorphous. Its role is equivalent to the role of theory and theoretical frameworks in science. Data are not collected at random. What is to be called data is already determined by some prior theoretical or quasi-theoretical exercise, implicit or explicit.

We have to recognize, therefore, that before the problem is further specified by the gathering of data about the problem, it is already powerfully constructed by two sets of limiting factors: the external constraints (although some of these may still be poorly understood) and the designer's cognitive capability in relation to that type of problem. It is quite likely that these latent limitations are already being explored right from the beginning, if the designer is conjecturing possible solutions, or at least approximations of solutions, in order to structure his understanding of the problem, and to test out its resistances. There is also a very practical reason why conjectures of approximate solutions should come early on. This is that a vast variety of design decisions cannot be taken—particularly those which involve other contributors—before the solution in principle is known.

As the designer collects and organizes the problem data, and data about constraints, his conjectures acquire sharper definition. Previously he was not able to test them out in a very specific way. Now he has an increasing fund of information against which to test them. He will also be using this information heuristically by using it in relation to his informal codes (see above p 254) by which abstract requirements are linked to built outcomes, and conjecturing further specifications within his roughly conjectured solutions. Information which has been used heuristically, can also be used to test the new conjectures. Conjecture and problem specification thus proceed side by side rather than in sequence. Moreover conjectures do not, on the whole, arise out of the information although it may contribute heuristically. By and large they come from the pre-existing cognitive capability—knowledge of the instrumental sets, solution types, and informal codes, and occasionally from right outside—an analogy perhaps, or a metaphor, or simply what is called inspiration. At least within this conceptualization of design we do not have to say that designers who use these last three types of source for conjecture are acting in a way that is markedly different from the architect with more modest ambitions. He has simply widened the scope of his conjectural field, sometimes moving right beyond the limits of the instrumental sets that are available.

When a conjectural approximation of a solution stands up to the test of the increasingly specific problem data (bearing in mind that it is always possible to collect more data and to produce more conjectures) a halt is called to both conjecturing and data gathering, and a solution in principle is agreed to exist. Further specification then takes place (i.e. further variety reduction) by completing a full design, and this is followed by a further refinement when the final production drawings are made. Unless the designer has great foresight, it is likely that further refinements will be made at the building stage.

We believe that this is more or less how design happens in most situations, and we believe moreover that it is as rational a process as is possible in the complex circumstances, not sub-rational because it is not 'systematic' and because so much depends on how the designer prestructures the problem. This outline model differs from the analysis–synthesis model (which we take to be the dominant notion in design method studies, hitherto) in several important ways. First, its core stratagem is conjecture—analysis rather than analysis–synthesis. Secondly the purpose of analysis is primarily to test conjectures rather than to optimize by logical or magical procedures. The notion of optimizing which architects believe they carry out can be easily contained within a conjecture–test psychology of design. Thirdly the solution in principle is allowed to exist at a much earlier stage than in the analysis–synthesis model. Fourth, the model shows the path of convergence on a unique solution without introducing notions like the optimization of information which, while attractive theoretically, are largely unlifelike and unworkable. Fifth, the model suggests *within its basic concepts* the possible origins of solutions in principle, a matter on which the design methodologists are notoriously silent or mysterious. Sixth, the model corresponds to the observed sequences of products of design, namely a set of descriptive documents of increasing refinement and specificity. Seventh, it recognizes implicitly that both information and conjectured solutions are inherently incomplete, but a stop has to be called somewhere. This is precisely equivalent to the situation in science. Eighth, and perhaps most important, the model emphasizes the importance of the designer's prestructuring of the problem, rather than denigrating it. It recognizes that architects approach—and should approach—design holistically and not piecemeal.

What does this have to say about research? We have already argued that presenting the 'results' of research in the form of packaged information or quantification tools does not seem to lead easily to better solutions. Perhaps the model will help to explain why. It is largely because unless research can influence designers at the stage of prestructuring the problem in order to understand it, then its influence on design will remain limited.

To explore this further, we might usefully examine the outcomes of research in terms of the four main types of elements which characterize the designer's field, namely instrumental sets, solution types, codes and information. It can be seen that much research of a purely technological kind (still by far the largest investment in building research) has its outcomes in terms of instrumental sets. Development work extends this into solution types by proposing exemplars. Research which aims to provide a method of checking design proposals against abstract requirements can be seen as a partial formalization of codes (partial because it is concerned with testing rather than generation and it is piecemeal). And research which has its outcome in the form of 'results', rather than a tool, falls into the field of information.

It can easily be seen that the first and last of these do not really help the designer to design. They normally increase the complication of the field and obscure its structure. Certainly they do not help the designer much at the stage of prestructuring the problem, and if they do so, it can only be in a haphazard way. The exemplars and prototypes that are the outcomes of development work certainly help the designer to prestructure his problem, but only if he proceeds in a largely imitative way. If the development is inadequate in any respect, it leads to a proliferation of these inadequacies. Over and above this, the prototype may be poorly understood, or badly adapted. Research in the third category is similarly unhelpful at the crucial stage of prestructuring. It may provide a means of eliminating errors at the design testing stage, given that the designer is able to use them properly, but we can hardly conceive of the designer being able to effectively utilize the full panoply of such techniques that would be required to cover all aspects of the design.

Of the four, only the development model can demonstrate to the designer new ways of prestructuring his problem. In spite of its disadvantages, its potential usefulness should not be underestimated. We could say that it suggests an organizational solution to the problem of linking research effectively with design. If research workers work with designers in producing experimental prototype solutions, which are intensively monitored and improved, then explained and publicized, then research itself benefits by becoming part of a dynamic process from which it can continuously learn and develop its concepts. In the past, development work in building has tended to lack both the deep involvement of research workers, and a properly developed monitoring function linked to a building programme. If both of these are provided for, there is at least an opportunity for sustained development over a period. By the quality and conviction of its examplars it can lead quite rapidly to a diffusion of real improvements in solutions.

On the other hand, the disadvantages of relying wholly on this fail-safe means of linking research with design are strong. The individual designer becomes severely constricted, problems of poor interpretation and debasement are likely to arise, creative innovation may be cut off or inhibited. Is there not some way in which research may help the designer to prestructure his problems more effectively without predetermining the solutions?

We believe there is, and that it lies in the notion of codes, the third element in the designer's field. Informal or implicit codes, we suggested, were used by the designer to link abstract functional requirements with instrumental sets, which no longer contained such codes. Taken together as a system, they constitute a kind of quasi-theory by which the designer structures his problem and finds a route through it—or through as much as is left of the problem after other external and internal constraints, including solution types, have had their say. Sometimes these codes are formalized and externalized in a rather pragmatic and programmatic way as 'architectural theory'. The influence and rate of diffusion of such externalizations is often very considerable. On occasion their impact is such as to have a marked effect on the development of instrumental sets.

The idea we are working towards, stated simply, is that research should aim (and is already beginning to aim) at the progressive reconstitution of the codes on a conceptual base by studies of people and their built environment which are oriented towards theory rather than 'results'. This is a complex and long-term aim, but it is entirely consistent with the normal impact of scientific work on human activities. The difference between a craft and a technology is not research results, but theory which brings structure and classification into phenomena, and allows the possible to emerge from an understanding of the actual. In any problem-solving activity, theory is the essential link between science and action. Without theory and its classificatory and route-finding possibilities, design is likely to remain, even in a field of endlessly proliferating scientific 'information', a kind of craft without continuity.

Here we come back to the reasons for optimism about architectural research. It seems to us that we are seeing the development of strong research programmes which are architectural in that they deal with broad bands of connected factors in design, and fundamental in that they are concerned with theories which actually relate to these levels of integration, rather than theories about isolated factors in environment. We would therefore like to try to explain what we see as the emerging structure of architectural research, why it is theoretical in a design as well as a scientific sense, and why it appears capable in the long run of affecting the ways in which problems are prestructured by designers.

We can best explain this by asking a question. What, in theoretical terms, is a building? On the grounds that buildings are not gratuitous but entirely purposeful objects, we would define a building as a realization of a number of social functions with an effect of ecological displacement. By specifying these functions and displacement effects in sufficiently abstract terms, we can formulate an adequate theoretical description of what a building is (such that anything which lacks one of them is not a building, and if an object is a building, it will fulfil all these functions whether by intention or as a by-product) and what its displacement effect is in terms of a four-function model. These are not true for all time, but are an historically accumulative set which define more or less what a building is at this point in time.

First, a building is a climate modifier, and within this broad concept it acts as a complex environmental filter between inside and outside, it has a displacement effect on external climate and ecology and it modifies, by increasing, decreasing, and specifying, the sensory inputs into the human organism.

Second, a building is a container of activities, and within this it both inhibits and facilitates activities, perhaps occasionally prompting them or determining them. It also locates behaviour, and in this sense can be seen as a modification of the total behaviour of society.

Third, a building is a symbolic and cultural object, not simply in terms of the intentions of the designer, but also in terms of the cognitive sets of those who encounter it. It has a similar displacement effect on the culture of society. We should note that a negatively cultural building is just as powerful a symbolic object as a positively (i.e. intentionally) cultural one.

Fourth, a building is an addition of value to raw materials (like all productive processes), and within this it is a capital investment, a maximization of scarce resources of material and manpower, and a use of resources over time. In the broader context of society it can be seen as a resource modifier.

In brief a building is a *climate modifier, a behaviour modifier, a cultural modifier, and a resource modifier*, the notion of 'modification' containing both the functional and displacement aspects.

Each of these functions can be conceived of separately as a people–thing relationship and each, in contrast to research oriented towards the 'atoms of environment' deals with a holistic set which constitutes *one way* of looking at a design problem. Each is capable of developing theory about people and their built environment. We would argue that research is gradually organizing itself within these foci as a set of interdependent, theory-oriented and largely structural studies, and that these are emerging as the fundamental disciplines of architectural research, and providing the base within which various disciplines become integrated and lose their identity.

It is notable, by the way, that the emphases implicit in this model shift architectural research right away from the study of procedures of design and into the study of buildings and their occupants, as well as away from 'results' and towards theory. We are beginning to look again at ends rather than means.

How will such research contribute to design? We have argued at the general level that it will progressively enable us to reconstitute codes from a theoretical base concerned with the relations between physical environments and those who experience them. We may add first that we conceive of this happening not in a positivistic and piecemeal way, but, because of the theoretical base, in a more holistic, non-deterministic and heuristic way. But this is too general a statement to be useful. We must specify further what we mean, and show why we can use this idea to escape from the idea of once-and-for-all 'knowledge' and allow for fundamental shifts in the theoretical bases by which we define 'knowledge' which will undoubtedly occur. We appeal again to the lessons of science.

In spite of periodic epistemological crises, paradigm switches and the progression and degeneration of research programmes, science continues to build its usefulness (as it has always done) on the strength of precise descriptions of the world. The theories on which these precise descriptions are based may be incomplete and even wrong, but they enable us to organize more and more of the world into useful cognitive schemes which, among other outcomes, enable us to conceive the possible out of a study of the actual. It might not be going too far to characterize the history of science as a series of immensely fertile delusions.

We do not therefore need to invoke the idea of 'knowledge' in order to propose that out of the notion of a building as a multi-functional object, and design as a multi-theoretical activity, we can begin to build up theory-based descriptions of the basic elements in design. These basic elements include ranges of activities, movements, perception-motivated actions, social intercourse patterns, spaces, and the environmental criteria that will satisfy a classified range of possible uses, coded and described in terms of the technologies which make them possible. Such a breakdown we might call a *base component classification* for environmental action, which would shift both in response to theoretical changes and also in response to changes in the environmental objectives of society. From the point of view of the designer, such classifications and code formalizations would not be deterministic or constitute a set to be specified in relation to problem information, but would constitute an extension of the designer's basic cognitive capability, and provide him with—and this is really the point about science—*a position of strength from which to make his conjectures*. In other words he would be using theories operationalised and specified as far as possible in terms of externalized codes, linking instrumental sets to human usage, as a basis for proposing his own further modifications to the environmental field.

The implications of this for the current formal structure of design activity—particularly those concerned with briefing, one-off user studies, and the designer's ability to re-interpret the 'client's requirements'—are enormous, and to examine them in detail would require another paper. To give one example, in the area of activity–space relations, we can foresee the possibility of moving from the 'activity–space fit' notion which is implicit in current practice, towards much more fundamental theories about the capability of certain types and configurations of space to contain an unpredictable variety of activities, perhaps with consequences for the idea of building types, and even for the size of cities. Such theories are not pseudo-deterministic ways of telling the designer what will be the outcome of his design, but strong and cumulatively developing bases for conjecturing possible futures.

If we are right in thinking that this is the underlying direction of the new lines in environmental research, then the notion of research simply as a service to design and the by-product of an eclectic variety of disciplines has to go by the board. Research is of course necessarily multi-disciplinary. In fact in the environmental field there appear to be no limits to the disciplines that could contribute to the advancement of the subject. But the contributions of the wider areas of science will only become effective through the integrative theories which will increasingly form the fundamental disciplines of environmental action itself, and these disciplines are not separate from design, but extensions of it in that their subject matter is design just as the subject matter of design is sets of instructions for building.

This is not a strange or unique arrangement. In fact it is very similar to science itself, seen in its broadest terms as one of the activities of society. Through science we continuously modify the world we live in and our understanding of it—that world and that understanding that it is the aim of science to study.

REFERENCES

Colquhoun, A. (1967), 'Typology and design method', *Arena* (June), pp. 11–14.

Keene, S.C. (1959), *Introduction to Meta-Mathematics*, North-Holland, Amsterdam.

Kelly, G. (1964), *The Psychology of Personal Constructs*, Wiley, New York.

Kuhn, T. (1962), *The Structure of Scientific Revolutions*, University of Chicago Press, Chicago.

Lakatos, I. and Musgrave, A. (eds) (1970), *Criticism and the Growth of Knowledge*, Cambridge University Press, Cambridge.

Levi-Strauss, C. (1966), *The Savage Mind*, Weidenfeld and Nicolson, London.

Nagel, F. and Newman, J.R. (1959), *Gödel's Proof*, Routledge and Kegan Paul, London.

Piaget, J. (1971), *Structuralism*, Routledge and Kegan Paul, London.

Popper, K.R. (1959), *The Logic of Scientific Discovery*, Hutchinson, London.

Popper, K.R. (1963), *Conjectures and Refutations*, Routledge and Kegan Paul, London.

Royal Institute of British Architects Research Committee (1970), 'Strategies for architectural research', *Architectural Research and Teaching*, **1**(1)(May), 3–5.

4.2 The Logic of Design

Lionel March

ATOMISM AND FALLIBILISM

In reviewing some of Alexander's work [in the first part of this paper, not included here] attention has been drawn to two fundamental aspects of his view. The first is his conception of a language, in Wittgenstein's sense, of design. The second is an atitude towards scientific method, which stems it seems from Popper. Now in their own fields both Wittgenstein and Popper are eminent philosophers. The difficulties of combining their two approaches as foundations for design theory will be apparent from Popper's strong opposition to Wittgenstein. The Vienna Circle's criterion of verifiability is opposed by Popper's criterion of falsifiability of empirical scientific systems (see Popper, 1961a; Wittgenstein, 1922). Wittgenstein puts his faith in the creator of a universe of irreducible atomic propositions which are true and which are the building blocks of all possible true structures. The truth of any structure follows from the truth of its atoms. The relations in which propositions stand to one another are internal. There is no need to set up relations between them. Their existence is an immediate result of the existence of the propositions.[1] It seems to us that this is precisely the attitude that Alexander adopts as a basis for a theory of design. The critical problem is to correlate this logically determined model with reality on the presumption that 'things are related to one another in the same way as the elements of the model'.[2] This is the task of the empirical sciences and it is this problem that Popper addresses. The doctrine of fallibilism that he proposes, and which Alexander adopts to tests his hypotheses, is an outcome of Popper's recognition of the demarcation between logic and mathematics on the one hand and the empirical sciences on the other. Popper

Originally published as part of 'The logic of design and the question of value' in March, L. J. (ed.) (1976), *The Architecture of Form*, Cambridge University Press, Cambridge. Reproduced by permission of Cambridge University Press.

draws attention 'to the invasion of metaphysics into the
scientific realm' caused by the positivists' wish

to admit, as scientific or legitimate, only those statements which are
reducible to elementary (or 'atomic') statements of experience—to
'judgment of perception', or 'atomic propositions', or 'protocol-
sentences', or what not. It is clear that the implied criterion of
demarcation is identical with the demand for an inductive logic. ...
And it is precisely over the problem of induction that this attempt to
solve the problem of demarcation comes to grief; positivists, in their
anxiety to annihilate metaphysics, annihilate natural science along
with it. For scientific laws, too, cannot be logically reduced to
elementary statements of experience.[3]

The philosophy of Karl Popper has had some influence on
modern architectural design theory. In the main its impact has
been pernicious, but this is as much the result of misunder-
standings as it is of Popper's own shortcomings. Just as Popper
draws a distinction between logic and empirical science, so too
must a distinction be made between these and *design*. To base
design theory on inappropriate paradigms of logic and science
is to make a bad mistake. Logic has interests in abstract forms.
Science investigates extant forms. Design initiates novel forms.
A scientific hypothesis is not the same thing as a design
hypothesis.[4] A logical proposition is not to be mistaken for a
design proposal. There has been much confusion over these
matters, hence the illusions about scientifically testable design
hypotheses and value-free proposals. Partly this is due to
Popper's restrictive view of logical forms, scientific hypotheses
and probability theory. Popper's theory 'stands directly
opposed to all attempts to operate with the ideas of inductive
logic'.[5]

'According to Popper's view, typical scientific hypotheses
are universal statements, that is to say, of the form "All *Y*'s are
Z" and are therefore falsified by one contrary instance. It is this
characteristic of falsifiability that makes them scientific, and
not any inductive support or any degree of probability
calculated from evidence.'[6] And finally Popper rejects, along
familiar lines, any subjective theory of probability in which
our 'rational beliefs' are not 'guided by an objective *frequency
statement*. This then is the information upon which our beliefs
depend'.[7] Yet in design, the chief mode of reasoning is
inductive in tenor, that is to say, synthetic rather than analytic:[8]
a good design hypothesis is chosen in the expectation that it will
succeed, not fail:[9] and since many designs are unique, probability theory, to be of any use, cannot be bounded by a frequentist
interpretation of probability 'in the long run', but must adopt
the subjectivist interpretation concerning 'degrees of belief'.[10]
Thus in design, in contradistinction to scientific discovery as
Popper would have it, the Popperian criteria must be stood on
their heads *in order to* sustain an approach which is rational.

THE LOGIC OF DESIGN

Here it is useful to turn to some logical investigations by the American philosopher Peirce in order to distinguish the nature of inductive reasoning and its relationship to other modes of plausible reasoning. In general, Peirce asserts: 'Let any human being have enough information and exert enough thought upon any question, and the result will be that he will arrive at a certain definite conclusion, which is the same that any other mind will make under sufficiently favourable circumstances'.[11]

Peirce (1923) takes the Aristotelian syllogism

$$x \text{ is } y; \quad y \text{ is } z: \quad \text{hence } x \text{ is } z$$

as typifying deductive or analytic reasoning, that is, the application of a general *rule* (y is z) to a particular *case* (x is y) to give a logically determined *result* (x is z). But, says Peirce, 'inductive or synthetic reasoning, being something more than the mere application of a general rule to a particular case, can never be reduced to this form'.

Having suggested that the syllogism is isomorphic to the transitivity axiom of a partial ordering,[12]

$$x \leqslant y \quad \text{and} \quad y \leqslant z \Rightarrow x \leqslant z,$$

Peirce goes on to develop two further modes of reasoning by permuting the three assertions (Figure 1). The two new modes are not logically determinate. Both are synthetic, involving leaps of the imagination in the sense of Whitehead's speculative reasoning.[13] Induction mirrors the reasoner's search for a law to account for regularities among phenomena and engenders new habits of thought.[14] Induction is the inference of the *rule* from the *case* and the results. Abduction, or what we shall later call productive reasoning,[15] Peirce's third mode, reflects the researcher's presumption that a certain phenomenon might exist to account for his observations given that a particular theory holds: abduction is the inference of a *case* from a *rule* and *result*.

Induction is where we generalise from a number of cases of which something is true, and infer that the same thing is true of a whole class. Or, where we find a certain thing to be true of a certain proportion of cases and infer it is true of the same proportion of the whole class. Abduction is where we find the some curious circumstance, which would be explained by the supposition that it was a case of a general rule, and thereupon adopt that supposition. Or, where we find that in certain respects two objects have strong resemblance, and infer that they resemble one another strongly in other respects.[16]

We usually conceive Nature to be perpetually making deduction in *Barabara* (the deductive syllogism). This is our natural and anthropomorphic metaphysics. We conceive that there are laws of Nature, which are her rules or major premises. We conceive that cases arise under these laws: these cases consist in the prediction, or the occurrence, of causes, which are the middle terms of the syllogisms.

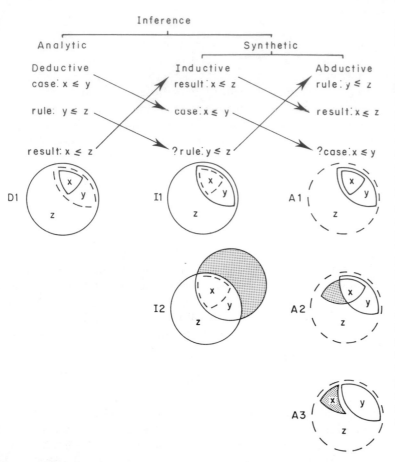

Figure 1
Peirce's three modes of inference.
There is one form of analytic
reasoning: the deductive, shown in
D1 as logically determined. There
are two forms of synthetic
reasoning: the inductive and the
productive. The hope in inductive
reasoning is to arrive at the
conclusion shown in I1. However,
there is no logical necessity for this
and the typical outcome must look
like I2 where the shaded part of y
indicates the amount by which the
rule $y \leq z$ is not met. Abductive
reasoning has three distinct
possibilities. In A1, as in the ideal
world of Sherlock Holmes, the
motive (rule) and the evidence
(results) conspire 'beyond all
reasonable doubt'—but without
logical certainty—to prove the
accused guilty (case). In A2, more
typically, there is a shadow of
doubt marked by the shaded part
of x suggesting the degree by
which $x \leq y$ is not supported. A3 is
yet another possibility. Here the
evidence and the motive simply do
not tie up: x, the shaded zone, is
disjoint from y.

And, finally, we conceive that the occurrence of these causes, by virtue
of the laws of Nature, results in *effects* which are the conclusions of the
syllogisms. Conceiving of nature in this way, we naturally conceive of
science as having three tasks — (1) the discovery of laws, which is
accomplished by induction; (2) the discovery of causes, which is
accomplished by hypothetic (abductive) inference; and (3) the prediction of effects, which is accomplished by deduction.

Deduction proves that something *must* be; induction shows that
something *actually* is operative; abduction merely suggests that
something *may be*.[17]

But whereas the major goal of scientific endeavour is to
establish general laws or theory, the prime objective of
designing is to realise a particular case or design. Both require
deduction, the quintessential mode of mathematical reasoning,
for analytical purposes. Yet science must employ inductive
reasoning in order to generalize, and design must use productive inference so as to particularize.[18] In attempting to draw
distinctions between induction and productive reasoning, or
for that matter science and design, nothing could be more
confusing than the fact that the word hypothesis has become its
own antonym. In science an hypothesis is a general principle
induced from particular events and observations, but in design

an hypothesis is a particular instance produced from a general notion and specific data. In science, hypothesis is commonly used to mean a tentative general statement about a class of cases (and this is its Popperian sense), but it originally meant a particular case of a general proposition[19] (and this is closer to what Peirce means by hypothesis). This semantic ambiguity accounts for widespread confusion and misunderstanding. It has also led to the use of the term induction to cover both varieties of Peirce's synthetic modes. Induction, for example, in decision theory is as often as not abduction.

For the purpose of developing a vocabulary for design theory the following terms will be used: the outcome of productive reasoning is a case which is called the design or *composition*—the latter in accord with traditional architectural theory;[20] the outcome of deductive reasoning is a *decomposition* which comprises the characteristics of the design that emerge from analysis of the whole composition—the whole is not merely the sum of these characteristics;[21] and the outcome of inductive reasoning is a *supposition*, a working rule of some generality—that is, an hypothesis in the scientific sense and more loosely, an idea, a theory, or in their modern usage a model, a type.[22] To rephrase Peirce's remarks above:

We conceive of rational designing as having three tasks—(1) the creation of a novel composition, which is accomplished by productive reasoning; (2) the prediction of performance characteristics, which is accomplished by deduction; and (3) the accumulation of habitual notions and established values, an evolving typology, which is accomplished by induction.

While it is from the collusion of specific needs and habitual notions that novelty is produced, it is productive reasoning which alone can frustrate the established order of habit and consequently inject new values. As Peirce writes: abduction, or as we have it production, 'is the only logical operation which introduces any new ideas; for induction does nothing but determine a value; and deduction merely evolves the necessary consequences of a pure hypothesis'.[23] Thus, production creates; deduction predicts; induction evaluates.

It is not the purpose of this paper to discuss in detail all three aspects of the design process, but to consider especially the problem of evaluation, which is seen to rest in the problem of inductive reasoning within the process. In these terms, however, rational design proceeds in this fashion:

01.1 From a preliminary statement of required characteristics and
01.2 a presupposition, or protomodel, we produce, or describe,
01.3 the first *design* proposal.
02.1 From design *suppositions* and theory and
02.2 the first *design* proposal

we deduce, or predict,

02.3 the expected performance *characteristics*.
03.1 From the performance *characteristics* and
03.2 the first *design* proposal
we induce, or evaluate,
03.3 other design possibilities, or *suppositions*.

The cycle then begins again:

11.1 from a revised statement of *characteristics* and
11.2 further, or refined, *suppositions*
we produce
11.3 the second *design* proposal,
and so on.

In this iterative procedure (Figure 2), it is assumed that certain characteristics are sought in a design to provide desired services, and that on the basis of previous knowledge and some general presuppositions or models of possibilities, a design proposal is put forward. Such a speculative design cannot be determined logically, because the mode of reasoning involved is essentially abductive. It can only be inferred conditionally upon our state of knowledge and available evidence. Deductive methods can then be used to predict measures of expected performance by the application of further models and theories to the particular design proposal. The design is usually

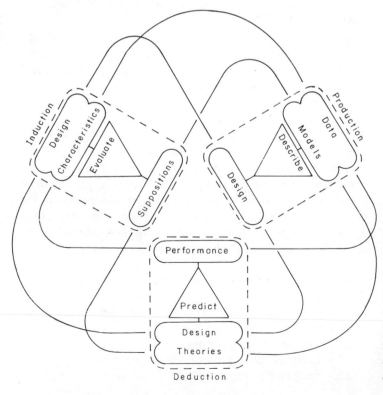

Figure 2
The PDI (production/ deduction/induction) model of the rational design process described in the text. The diagram suggests a cyclic, iterative procedure PDIPDIPD... and so on, with constant refinements and redefinitions being made of characteristics, design, and suppositions as the composition evolves. In fact the model is envisaged as representing a critical, learning process in that statements inferred at later stages may be used to modify those used in earlier stages and thus to stimulate other paths of exploration. For this reason no arrows are shown along these paths, although the general direction of argument is clockwise.

represented by a set of drawings, and its analysis may be little more than eye-balling and an intuitive feel for how it will work out in practice. Nowadays, however, computer representation is possible so that a more formal prediction of expected performance can be caried out using mathematical models. The mode of prediction, in the sense used here, is essentially deductive: for example, if the design is like this, and the laws of light are like that, then it follows that the lighting in this space will be so and so. Deductive inference is determinate, but the premises may involve probabilistic statements, in which case the conclusion will also be probabilistic, not because of any uncertainties in drawing the inference itself but because of the nature of the premises. It should be noted that at this stage further theories may well have been brought forward and that new characteristics will have been generated which may critically augment the original set in the previous productive phase. In the inductive stage the design and its expected characteristics are used to infer new generalizations and suppositions. Induction criticizes the original presuppositions in the productive phase and provides more discriminating models for the next round of the cycle. But it does more. We have said that induction is evaluative. What induction does is to take a specific instance and to infer a general setting for it. In itself a design, or rather the set of pertinent characteristics by which it is perceived, has no value. It assumes relative value through comparison with other designs both executed and entertained as well as the environment as a whole. Indeed, evaluation assumes that suppositions about worth, preference, desirability, or utility can be inferred. It is these suppositions that form the substance of the productive phase of designing. That is to say, the models required to produce design alterna-tives are value-laden. It follows that value theory is the essential foundation of any rational theory of design. And it is to decision theory and micro-economics that we shall need to turn for guidance in its development.

Hawkes (1976) points out the role of abductive reasoning in the design process: 'this is simply that there is, at any point of time, a generally held notion about the nature of a good solution to any recurrent building design problem and that it is this notion which frequently inspires the initial hypothesis [in the sense of design]'. Such a notion, or presupposition, is value-laden. Those models are chosen which are recognized as worthy. But, with respect to the design of the modern office building, Hawkes shows how a new design habit becomes entrenched and comes to dominate previous notions. At the same time there is a tendency for the current stereotype, or general notion, to become too highly specialized so that its progressive evolution is jeopardized:

The answer appears to lie in the realisation that the current stereotype does not supersede all others. There is, in fact, a *store* of accumulated experience which contains all previous solutions and which will be enlarged in the future with the addition of new examples inspired by changing building technology, organisational ideas and physical, social and cultural environments. This view demands a return to earlier stereotypes to see what they might offer when modified to exploit developments in technology which have occurred since their day. A healthy situation would be one in which solutions with a high dependence on technology could co-exist with others which achieved their goals by simpler means.[24]

The survival of stereotypes in the past has been a matter of trial and error in practice: the fittest have survived as evidence of their utility. Such evidence is usually censored by a designer's judgment, collected through his experience, and in time becomes part of his intuitive response. If internalized personal judgment, experience, and intuition alone are relied upon, the three modes of the PDI-model become inextricably entangled and no powerfully sustained use of collective, scientific knowledge is possible. Design will remain more or less personalistic and a matter of opinion, albeit professional. If the design process is externalized and made public, as it evidently must be, for team work to be fully effective, then the three stages of the PDI-model are worth making explicit so that as much scientific knowledge can be brought to bear on the problem as seems appropriate. In this externalized process it is feasible to experiment with artificial evolution within the design laboratory using simulated designs and environments. New, synthetically derived stereotypes may emerge, and old ones may be given new potential without having to wait for practical exemplification. Design comes to depend less on a single occasion of inspiration, more on an evolutionary history, greatly accelerated as this iterative procedure can now be—a prospect opened up by recent advances in computer representation.

Notes

1 Wittgenstein (1922), prop. 5, 1 ff.
2 Wittgenstein (1922), prop. 2, 1 ff.
3 Popper (1961a), pp. 35–6.
4 Langer (1953), see Chapter 1, 'The study of forms', 32.a. See below, p.268.
5 Popper (1961b), p. 30.
6 Hesse (1963), especially Chapter 3, 'The logic of analogy'.
7 Popper (1961a), p. 211.
8 'Inductive logic is more suited for engineering design than is deductive logic. The characteristic feature of the problem is that the designer does not know *everything*, yet he does know *something*. He cannot be certain about what he ought to do, but he does have many clues. Problems of inductive logic always leave a residue of doubt in the mind. Most people prefer to find a solution by deductive logic because then they can be sure they are right. Often engineers will pretend that a problem is capable of being solved by deduction, simply because they have a greater facility

with this kind of reasoning and because the answer leaves them with a greater sense of satisfaction' (Tribus, 1969, p. 4). Also see below, p.267, for a distinction between two modes of synthetic inference: induction and abduction. We argue there that it is the latter which plays the key role in designing.

9 Alexander (1964) identifies the design requirements 'from a negative point of view, as potential misfits' between form and context. The synthesis of form sought by the designer involves the successful conversion of these misfits into a harmonious ensemble. Alexander himself has never suggested that a 'design hypothesis'— a project taken as a whole—is testable in Popper's sense.

There is another aspect of success and failure. When a new design is being developed, 'there is very little information on which to base a prediction as to its life. After a few units are put into the field, failure data begin to come in. The data should, if at all possible, be examined in detail for the causes of failure and these causes used for redesign' (Tribus, 1969, p. 430). It is remarkable how rarely this is done in architectural design where units are frequently repeated. For unique designs such an approach is obviously not possible, but see the next note.

10 The distinction between the 'frequentist' and the 'subjective' interpretations of probability might seem small, 'but it is the key to the power of decision analysis. The objectivist (frequentist) requires repeatability of phenomena under essentially unchanged situations to make what he would consider to be meaningful inferences. The subjectivist can accept any amount of data, including none, and still apply logic to the decision. The objectivist was able to survive and even flourish, when the main problems of inference arose in areas such as agriculture that provide large amounts of cheap data. Today, when decisions regarding space programmes must be based on a single launch of a one hundred million dollar rocket, the ability of the subjectivist to apply logic to one-of-a-kind situations has become indispensable' (Howard, 1968).

11 Peirce (1871). Peirce's work is most conveniently summarized in Feibleman (1970). There is also a popular exposition in Peirce (1923). Peirce, like his fellow American the physicist William Gibbs, seems to have been treated like a country-cousin by his more urbane European peers and their Yankee courtiers (especially from the East Coast academic establishment). Peirce, as early as 1878, appears to have been the first to have proposed a logarithmic expression for a measure of belief—akin to information entropy in modern usage—although his opinions on probability theory followed the English logician Venn in rejecting the subjective viewpoint of Laplace and Bayes in favour of a strictly 'frequentist' one. However, in a series of important contributions, Jaynes (1963, 1967, 1968) has shown how Gibbs had effectively used the entropy concept to overcome objections to the subjectivist approach. 'Once we had Shannon's theorem establishing the uniqueness of entropy as an "information measure", it was clear', writes Jaynes, 'that Gibb's procedure was an example of a general method for inductive inference, whose applicability is in no way restricted to equilibrium thermodynamics or to physics.' And then, almost paraphrasing Peirce: 'Two observers, given the same set of propositions to reason about and the same prior evidence, must assign the same probabilities.' The crucial problem is how to assign a unique prior probability. 'Fortunately, we are not without clues as to how this uniqueness problem might be solved. The principle of maximum entropy (i.e. the prior probability assign-

ment should be the one with the maximum entropy consistent with the prior knowledge) gives a definite rule for setting up priors. The rule is impersonal and has an evident "intuitive appeal" as the distribution which "assumed the least" about the unknown parameters', and accords with 'an ancient principle of wisdom— that one ought to acknowledge frankly the full extent of his ignorance'. This approach also accords with the spirit of Peirce's own remarks concerning human knowledge.

12 See March (1976).

13 See Martin (1967), who refers to Whitehead's essay *The Function of Reason* (1929).

14 See Peirce (1923), p. 15, 'the belief in a rule is a habit. That a habit is a rule active in us, is evident. Induction, therefore, is the logical formula which expresses the physiological process of formation of a habit' or in the design context, perhaps, the establishment of a type.

15 Peirce did not use the term productive: he used at different times the terms abductive, retroductive, presumptive, hypothetic. In the design context our choice of term seems more telling and natural.

16 Peirce (1923), p. 135.

17 Hartshorne and Weiss (1931–5), vol. 2, para. 173, and Vol. 5, para. 171, respectively.

18 It should be noted here that, whereas inductively derived scientific suppositions tend to be *universal*, productively derived designs are *existential*. Thus, while in science it might behove the scientist to search for a falsifying case; in designing, the problem—as any designer knows—is to find *at least one* reasonable design solution.

19 The leading, but obsolete, definition in the *Shorter Oxford English Dictionary*.

20 Herein is the reason for our preference for the term production rather than abduction. We *produce* a composition, and a design composition is a *product*.

21 The term decomposition has established itself in architectural design-theory literature through the pioneering work of Alexander (1965). See Part 4, 'Identification of problem structure', of Moore (1970) for a review of decomposition techniques. The major criticism concerning these techniques is the implicit assumption that the structure of the problem can be derived from the manipulation of atomic statements. The difference between this approach and that adopted here is contained in the apposition: 'the atoms (or characteristics) are seen as things which can be understood as parts of the whole (or composite)'—our view, and 'understanding the structure of the whole (or composite) as a thing which is assembled from the atoms (or parts)'. See Atkin *et al.* (1972) for an approach to structure more in line with ours using the theory of simplicial complexes.

22 See Echenique, 'Models: a discussion', in Martin and March (1972), pp. 164–74.

23 Hartshorne and Weiss (1931–5), vol. 6, para. 475, which also contains the aphorism: 'Deduction explicates: induction evaluates.'

24 See Willoughby (1976). An evolutionary approach to design suggests parallels with the biological concepts of genotype and phenotype in which the former relates to the 'deep' structure of a whole evolving population, and the latter to the 'surface' structure of an individual instance of that population resulting from the interaction of the genotype with a succession of environments, or as Dobzhansky (1960) puts it: 'The appearance, the structure, and the functional state which the body has at a given moment is its phenotype. The phenotype of a person changes continuously

through life as his development proceeds. The changes in the manifestation of heredity in the phenotype contrast with the relative stability of the genotype.' Hawkes's stereotype and our supposition, as a type, are by analogy genotypical; a specific design proposal is phenotypical, a particular transform of what might be called a 'kernel' form. Most design theory concentrates on the development of the phenotypical design and we are largely concerned with this here, but we also argue that it is now conceivable, through the artificial evolutionary processes made possible by modern computer simulation, to design (evolve) new architectural genotypes. This should be distinguished from the traditional creation of a new type—an exemplary phenotype—which influences future design through the 'contagion of success'.

REFERENCES

Alexander, C. (1964), *Notes on the Synthesis of Form*, Harvard University Press, Cambridge, Mass.

Alexander, C. (1965), 'A city is not a tree', *Architectural Forum*, April, pp. 58–62, and May, pp. 68–71. Reprinted in *Design*, February 1966, pp. 46–55.

Atkin, R.H., Mancini, V., and Johnson, J. (1972), *Urban Structure Research Project: Report 1*, Department of Mathematics, University of Essex, Colchester.

Dobzhansky, T. (1960), *The Biological Basis of Human Freedom*, Columbia University Press, New York.

Feibleman, J.K. (1970), *An Introduction to the Philosophy of Charles S. Peirce*, MIT Press, Cambridge, Mass.

Hartshorne, C., and Weiss, P. (eds) (1931–5), *Collected Papers of Charles Saunders Peirce*, Harvard University Press, Cambridge, Mass.

Hawkes, D. (1976), 'Types, norms and habit in environmental design', in March, L.J. (ed.), *The Architecture of Form*, Cambridge University Press, Cambridge.

Hesse, M.B. (1963), *Models and Analogies in Science*, Sheed and Ward, London.

Howard, R.A. (1968), 'The foundations of decision analysis', *IEEE Trans. on Systems Science and Cybernetics*, **4** (3), 211–19.

Jaynes, E.T. (1963), 'New engineering applications of theory', in Bogdanoff, J.L., and Kozin, F. (eds), *Symposium on Engineering Applications of Random Function Theory and Probability*, Wiley, New York.

Jaynes, E.T. (1967), 'Foundations of probability theory and statistical mechanics', in Bunge, M. (ed.), *Delaware Seminar in the Foundations of Physics*, Springer-Verlag, Berlin.

Jaynes, E.T. (1968), 'Prior probabilities', *IEEE Trans. on Systems Science and Cybernetics*, **4** (3), 227–41.

Langer, S.K. (1953), *An Introduction to Symbolic Logic*, Dover, New York (first published 1937).

March, L.J. (1976), 'A Boolean description of a class of built forms', in March, L.J. (ed.), *The Architecture of Form*, Cambridge University Press, Cambridge.

Martin, L. (1967), 'Architects' approach to architecture', *RIBA Journal*, May.

Martin, L., and March, L.J. (eds) (1972), *Urban Space and Structures*, Cambridge University Press, Cambridge.

Moore, G.T. (ed.) (1970), *Emerging Methods in Environmental Design and Planning*, MIT Press, Cambridge, Mass.

Peirce, C.V. (1871), 'The works of George Berkeley', *North American Review*, **113**, 449–72.

Peirce, C.S. (1923), *Chance, Love and Logic*, Kegan Paul, London.

Popper, K. (1961a), *The Logic of Scientific Discovery*, Science Editions, New York.

Popper, K. (1961b), *The Poverty of Historicism*, Routledge and Kegan Paul, London.

Tribus, M. (1969), *Rational Descriptions, Decisions and Designs*, Pergamon Press, Oxford.

Whitehead, A.N. (1929), *The Function of Reason*, Princeton University Press, Princeton, New Jersey.

Willoughby, T. (1976), 'Searching for a Good Design Solution', in March, L.J. (ed.), *The Architecture of Form*, Cambridge University Press, Cambridge.

Wittgenstein, L. (1922), *Tractatus Logico-Philosophicus*, Routledge and Kegan Paul, London.

4.3 Design and Theory Building

Geoffrey Broadbent

It is a truism to say that the histories of design have mostly been written as catalogues of changing styles: Egyptian, Greek, Early Christian, Byzantine, Romanesque, Gothic, Renaissance, and so on. Obviously design has changed over the years yet the founding fathers of modern architecture and design, such as Le Corbusier (1923), Gropius (1956), and others, believed that whilst such stylistic changes may have been needed in the past they would be quite unnecessary in the future. The proper definition of each design problem and its thorough analysis—using new techniques—would simply eliminate such changes. These founding fathers of modern design wanted to establish standards from which optimum design solutions could be derived, in which a concept of style would be quite irrelevant. They even started to establish such standards at their CIAM congresses, whilst the methods of analysis they sought became increasingly available out of ergonomics, operational research, systems analysis, and so on, greatly encouraged of course, by the advent of the computer (see my *Design in Architecture*, 1973). How strange that the forms they pioneered, whether in urban planning, architecture, furniture, or any other kind of design developed by the use of those analytical methods, should have been some of the most short-lived (perhaps 50 years) in the entire history of designing!

They have led to the wholesale destruction of small-scale, humane environments and their replacement by bleak, windswept piazzas and urban motorways, lined with the dimmest, greyest, and most soul-destroying of high-rise apartment slabs. In other words, we can now detect faults in the 'theories' of Gropius and Le Corbusier equivalent to—but obviously

Originally published in *Design Methods and Theories*, **13** (3/4) (1979), 103–7. Reproduced by permission of the Design Methods Group.

different from—those which they detected in the 'theories' of
their predecessors. But that is the nature of human affairs, and
the best description of the processes by which it occurs still is
Thomas Kuhn's in his *The Structure of Scientific Revolutions*
(1962). Kuhn believes that science itself—that most objective of
human affairs—actually changes from time to time, in a way
which seems almost like fashion. He believes that at any
moment in time, all the 'normal' scientists working in a
particular field share certain ideas, ways of working, and so on
to which he gives the collective—and somewhat ambiguous—
name of 'paradigm'.

He sees the majority of scientists as sound and unadventur-
ous, working on the basis of other people's achievements, as
described in the textbooks from which they learned their
science. The 'normal' scientist will be content to work within
the going paradigm, to solve the 'puzzles' embodied within it.
They will work on the same kinds of problems, use the same
kinds of apparatus, write up their work according to established conventions, using numbers, where possible, instead of
words or pictures. Indeed, they will have to if their papers are
to be accepted by respectable journals, conference organizers,
and so on. The paradigm, in this sense, will be a set of social
pressures, forcing the scientists to work in certain ways.

But the more imaginative, the more intelligent, the more
adventurous of them will realize that the going paradigm is by
no means the final solution to problems in their particular field.
They will detect flaws and anomalies within it and eventually a
Newton, an Einstein, or a Heisenberg will come along with a
new set of ideas so powerful, so unprecedented, yet so
convincing, that with some exceptions (who eventually will die
off) the 'normal' scientists will be 'converted' to this new point
of view. They will adopt it as the basis for their own work,
which will consist of exploring aspects of the paradigm which
are incompletely defined even in the new formulation. Similar
mechanisms obviously are at work when 'normal' designers are
'converted' to the 'theories' of a Gropius or Le Corbusier.

Unfortunately, as Masterman shows (1970), Kuhn uses the
word 'paradigm' in no less than twenty-one ways within his
basic text (1962)—some of which have been quoted above. She
groups them into three coherent sets including the beliefs,
myths, ways of seeing, standards, metaphysical speculations,
and organizing principles which she calls *metaparadigms*. Then
she identifies a set of *sociological* paradigms—relating to the
ways in which scientific achievements are recognized by the
community and so on. Kuhn (1970a) accepts Masterman's
criticisms, suggesting that if he were to write the *Structures*
again, he would concentrate specifically on the social pressures
by which the members of a particular scientific community are
bound together. Kuhn then goes on to suggest his own way of

ordering his paradigm within what he calls a 'disciplinary matrix'. Within this Kuhn's scientists will share certain *symbolic generalizations*, that is, the laws of nature which they *must* operate, the equations, definitions and generalizations which they all have agreed to use. Also they will be bound by a set of *shared commitments*, including beliefs in particular models or ways of describing the world, similar metaphors, or analogues in describing what they do and so on.

Kuhn's scientists also share a set of values: attitudes to the need for objectivity, accuracy, and consistency, expressed in mathematical terms. They share certain attitudes towards the social responsiblity of the scientist and indeed a whole set of shared values which determine their group behaviour. And finally Kuhn's scientists share certain exemplars—textbook solutions to various problems which every 'normal' scientist working within the paradigm will know and use as necessary.

There still remains the question, of course, as to how far Kuhn's idea of paradigm *can* be applied in other fields such as design. He says (1970b):

A number of those who have taken pleasure from it [*The Structure of Scientific Revolutions*] have done so less because it illuminates science than because they read its main thesis as applicable to many other fields...

I see what they mean and would not like to discourage their attempts to extend the position, but their reaction has nevertheless puzzled me. To the extent that the book portrays scientific development as a succession of tradition-bound periods punctuated by non-cumulative breaks, its theses are undoubtedly of wide applicability. But they should be, for they are borrowed from other fields. Historians of literature, of music, of the arts, of political development, and of many other human activities have long described their subjects in the same way.

So far, so obvious, but his crucial point is this:

I suspect, for example, that some of the notorious difficulties surrounding the notion of style in the arts may vanish if paintings can be seen as modelled on one another rather than produced in conformity to some abstracted canons of style.

Designers obviously are subject to group pressures of exactly the kind which Kuhn describes for his scientists. They receive similar educations (according to where and when they study), read the same textbooks, use the same techniques, read the same journals, attend the same conferences, have their own networks for contact and correspondence. Designers, too, are judged by their peers according to the number of times their work is published in the journals and the places they are invited to speak. Just like Kuhn's scientists, we too have our citation indices.

Kuhn's reformed paradigm actually can be stretched fairly easily to cover design; we certainly have our equivalents for its four constituent parts. Let us propose that the paradigm for design at a particular time includes:

(1) technical knowledge—including the equations, laws of nature, and so on by which Kuhn's scientists are bound;
(2) professional skills—which, in this connection, probably will include the ways of thinking by drawing, model-making, and so on equivalent to the analogues, metaphors, and 'models' by which Kuhn's scientists describe and understand what *they* are doing;
(3) ideologies—equivalent to the *values* shared by Kuhn's scientists;
(4) examples—as presented by such form-givers as Le Corbusier—published in journals and books and exactly equivalent to Kuhn's exemplars.

If the hardest-line scientists work within paradigms, then that must be the nature of human affairs. Designers, obviously, cannot be more rigorous than that so the notions of style and stylistic change after all, *are* fundamental to what designers do.

Does this mean that for ever more we shall be subject to the creative whims of those who will generate new architectural paradigms? In a literal sense it does. New Le Corbusiers *will* present new forms. Initially they will meet with resistance but some of them will catch on. But the paradigm will shift only if it is consistent with the professional skills and ideologies of the 'normal' designer. The trouble with the paradigm shifts of fifty years or so ago is that they were in conflict with the laws of nature.

To take the obvious example, the steel and glass curtain-walled office building of the mid-twentieth century. Gropius, Le Corbusier, Mies van der Rohe, and others presented the exemplars from which this particular form was derived. The form itself obviously came to symbolize business efficiency in the brave new world of the future—that was its ideological value, whilst the task of designing it obviously fell within the professional skills of currently practising architects, engineers, and associated professionals. But those who designed it completely ignored those laws of nature which determine the behaviour of glass, solar energy, noise, and so on. A proper understanding of these things would have led to a rejection of the paradigm. Designers in the future obviously will have to be aware of these things if they are to avoid equivalent faults in the new paradigms which replace it. And that, obviously, will depend on contributions from theory.

It so happens that certain pioneers of modern design, such as Hannes Meyer, whilst by no means seeing design as a matter of establishing standards and repeating exemplars, at least felt that theory from anatomy, physiology, psychology, and the social sciences would greatly increase the possibilities of generating 'good' design (see Schnaidt, 1965). Indeed, Meyer introduced such things into the design curriculum at the Bauhaus in 1927–30 for the first time in any design school, whilst his initiative was taken up by Maldonado (at Ulm, 1956–72). Sir Leslie Martin and (Lord) Llewellyn-Davies persuaded the RIBA at their influential conference on architectural education, held at Oxford in 1958, that architectural education in Britain also should be founded on theory, rather than on the crafts of current practice (see also Llewellyn-Davies, 1961). The problem, of course, is to know what kinds of theories are relevant, or even possible in design. 'Theory' as a concept is difficult enough in science itself, in which definitions of theory, not to mention the rigour with which theories are developed and sustained, vary so much from field to field. To take first of all the question of rigour, most philosophers of science sooner or later refer to planetary theory as the paradigm of what theories are and how they should be built. We can, in fact, use it to define what a theory *is* and admit that any so-called 'theory' actually is a *theory*, only in so far as it shares the salient characteristics of that theory.

The earliest planetary 'theory' seems to have emerged in Babylonian times, when it was first observed that whilst the majority of stars seemed to rotate in a synchronized manner around the heavens, a small number constantly were changing in their relative positions. These were observed, named, and their movements recorded; tables were then prepared which enabled the Babylonian astronomers to *predict* their movements into the foreseeable future. That had certain practical advantages; in navigation, for instance, and once events such as eclipses could be predicted, they need no longer be considered mysterious, to be held in awe and dread.

The Babylonian astronomers, of course, had their explanations for planetary motion; they thought of the earth as the centre of the universe, with the sun, moon, visible planets, and stars revolving around it. Neither these explanations nor the Babylonian predictions were particularly accurate, and a number of rival theories emerged in Classical times, including some which placed the sun at the centre of the universe. Accepted classical theory was summarized by Ptolemy of Alexandria in the *Almagense* (second century A.D.). Ptolemy thought that the earth was fixed and that the sun, moon, and planets described eternal circles around it. Data available even then, however, showed that none of the planets described an exact circle around the earth—Ptolemy explained the observed discrepancies by suggesting that they moved in *epicycles*, small rotations at the circumference of their main orbits.

This is not the place to develop a comprehensive history of planetary theory but we ought to note that Ptolemy's views held sway until the early years of the sixteenth century when Copernicus, dismissing Ptolemy's epicycles as far too facile, delved back into the forgotten rival theories and, after much persuasion, announced in 1510:

The centre of the earth is not the centre of the universe, for there is not just one centre for all the celestial orbits. All orbits encircle the sun, as if the sun were at the centre of all; hence the centre of the universe lies near the sun.... All movement that is visible in the heavens of fixed stars is not so *per se*, but as seen from the earth.

Copernicus was wrong in detail, but his explanation fitted the observed facts much more closely than Ptolemy's ever had. Although telescopes were not to be invented for another hundred years, increasingly accurate observations became available from Tycho Brahe's observatory towards the end of the sixteenth century. Brahe's instruments, and his data, were inherited in 1610 by Johan Kepler who, after covering some 9000 folio sheets with his tiny figures, finally concluded that Copernicus's actually fitted the observed facts much more closely than Ptolemy's.

But there were still discrepancies and by 1609 Kepler was suggesting that these could be diminished, at least in the case of Mars, by conceiving its orbit as elliptical rather than circular. A year later, in 1610, Galileo brought the first telescopes into astronomical use (he later sent one to Kepler) and thus increasingly accurate observations could be made. Galileo too was convinced by the Copernican explanation and pronounced it, in the *Dialogo sopre i due massimi sistemi del Mondo* of 1632, to be not just another hypothesis but the proven truth. Galileo, as everyone knows, was subject to the forces of social, political, and religious reaction which perpetuated the Ptolemaic paradigm for a little while longer, but already by the mid-seventeenth century an explanation of planetary motion, with predictive powers, was known, which essentially is that which we use today—the planets revolve around the sun in elliptical orbits.

Further details, certainly, have since been added to planetary theory and its predictive powers now are such that given any stated time in the near or distant future, we can predict, with something approaching certainty, where each of the planets will be, what relationships they will have to each other, and so on. The model is so powerful, in fact, that shortly after the (visual) discovery of Uranus in 1781 certain discrepancies were noted in its orbit which could only be explained by the existence of a further planet. The position of that planet was predicted mathematically; it was located by telescope in 1846 and named Pluto. Perturbations were also detected in Pluto's orbit and once the implications of these had been worked out, a further planet, Neptune, was finally located and discovered in 1930. The existence of many further objects orbiting the sun has been predicted mathematically, but so so far they have eluded visual observation.

Astronomy therefore presents a prototype, not just for the physical sciences, but for all science: decide that certain phenomena are of interest and try to explain them; collect data on whatever is measurable concerning those phenomena and note the degree to which one's explanations match the available data; if there are discrepancies, modify one's explanation accordingly; if it cannot be made to 'fit' the observed facts, then look for another explanation. Obviously with the planets measurements can be taken at intervals over time and given their positions at any one time one can predict—according to one's model—where they *ought* to be at all other times. One can then check discrepancies between one's predictions and 'what actually happened'. If discrepancies are revealed then one's model can be modified, and 'run' again to see if in this modified form it offers better predictions. If it fails then one might try further modifications or abandon it altogether and think up 'better' explanations. Eventually—as in the case of planetary theory—one hopes to finish up with an explanation, a model which is so good that it really does predict—at the desired level of accuracy—all the future states that the phenomena under discussion are ever going to be in.

The predictive nature of theory was stated with exemplary clarity by the French mathematician Pierre Simon de la Place, who writing in 1812 said:

We ought then to regard the present state of the universe as the effect of its anterior state and as the cause of that which follows. Given for one instant an intelligence which could comprehend all the forces by which nature is animated and the respective situation for the beings who compose it—an intelligence sufficiently vast to submit these data to analysis—it would embrace in the same formula the movements of the greatest bodies of the universe and those of the lightest atom; for it, nothing would be uncertain and the future, as the past, would be present to its eyes.

Throughout the build-up of planetary theory, neither man's observations of the planets—however inaccurate—nor his explanations of their motion—however false—had the slightest effect on how they actually moved.

If planetary theory is *the* exemplar of theory, then only a small proportion of those things to which the word 'theory' attached really *are* theories by our strict definition of the term, and even of those which pass this particular test, only some obviously are 'good' whilst others most certainly are 'bad'. So how do we distinguish between them? Karl Popper attempted to do that in 1921–2 with considerable success, thus laying the foundations for much of his later work (1963).

Popper, in fact, tried to lay down criteria for distinguishing between real science and pseudo-science, starting with the observation that Einstein's theory of relativity was different in kind, say, from Marx's theory of history, Freud's psychoanalysis, or Adler's 'individual psychology'. In seeking to explain the differences he found they were rather subtle. Einstein's theory was couched in mathematical terms and thus seemed more exact than the others which were presented verbally, but that was by no means the root of his unease. He was not even sure if Einstein's theory of gravitation could be *true*, whilst the truths of Marxism, psychoanalysis, and individual psychology certainly could be observed wherever he looked for evidence. Popper's friends believed that the beauty of these theories lay precisely in their explanatory power, for given the necessary commitment to their essential truths—which depended initially on a form of intellectual 'conversion'—a paradigm shift in Kuhn's terms—they could see confirming instances everywhere. The world was 'full of verifications'. Sceptics like Popper who refused to see them did so 'either because it was against their class interests, or because of their repressions which were still "un-analysed" and crying aloud for treatment'. But for Popper, it was precisely this fact 'that they always fitted, that they were always confirmed—which in the eyes of their admirers constituted the strongest argument in favour of these theories … this apparent strength was in fact their weakness'.

Einstein's theory, by contrast, only applied in specific and clearly defined cases. He predicted that light would be deflected by the gravitational effects of large bodies—a prediction which in itself was quite alien to most people's expectations. It was, quite simply, anti-common sense. As a theory therefore it had been at considerable risk until Eddington conducted a series of observations in which the deflection actually was seen to occur. If Eddington had failed to observe such deflections then the theory would have remained unproven. Or, more particularly if he had shown that Einstein's predicted gravitational effects were absent, then the theory, quite specifically, would have been refuted. But Eddington's observations on the contrary failed to confirm 'common sense' and Einstein's theory survived. All the theory-building, according to Popper, is a matter of generating *conjectures* and submitting them to rigorous *refutation* procedures.

Popper attempted therefore to write down the factors which characterize a 'good' theory. These may be paraphrased as follows:

(1) It is easy to verify any theory once we have set it up if we simply look for verification.

(2) Verifications of this kind are only worth having if the theory itself initially was *at risk*, if a common sense view—'unenlightened by the theory in question'—would have predicted something quite different.

(3) Instead of allowing almost anything to happen, and providing possible interpretations for it, a 'good' scientific theory will merely forbid certain things.

(4) In setting up a good theory it will be possible to conceive of certain events which almost certainly could refute it. Any theory which is not refutable in this way is non-scientific.

(5) The only genuine tests of a theory are those which attempt to *refute* or to falsify it. The degree to which theories are refutable varies greatly; some are more exposed to refutation than others, being at greater risk, they are therefore better theories.

(6) If a genuine attempt to refute a theory fails, and serves merely to confirm it, that confirmation is admissible; other kinds of 'confirmation' which have actually been sought are simply not worth having.

(7) Once a genuine theory has been tested and found wanting, its admirers may introduce further assumptions or interpretations which seem to salvage it. They may succeed, but in doing so they risk the lowering of its scientific status.

So, for Popper *'the criterion of the scientific status of a theory is its falsifiability, or refutability, or testability'*.

Popper points out that neither the writings of Adler nor those of Freud can meet his criteria. They were simply not testable, because there was no conceivable human behaviour which could possibly contradict them. Marx's theory of history was in a slightly different category, as Popper says (1963):

In some of its earlier formulations (for example Marx's analysis of the character of the 'coming social revolution') [their] predictions were testable, and in fact falsified. Yet instead of accepting the refutations the followers of Marx re-interpreted both the theory and the evidence in order to make them agree. In this way they rescued the theory from refutation, but they did so at the price of adopting a device which made it irrefutable. They thus gave a 'conventional twist' to the theory; and by this stratagem they destroyed its much advertised claim to scientific status.

In the case of the two psychoanalytical theories, however, Popper is careful to point out that their non-theoretical status in no way diminishes their power. As he says:

This does not mean that Freud and Adler were not seeing things correctly: I personally do not doubt that much of what they say is of considerable importance, and may well play its part one day in a psychological science which is testable.

He is less generous to Marx although not, one suspects, because the latter was unsuccessful. On the contrary, his objections to Marx arise from the fact that the latter *achieved* a social revolution of which he, Popper, disapproves. One could argue in fact that Marxism was successful in transforming the social system for one half of the world's population *because*, at base, it was not a scientific theory.

There are fundamental problems in any case about *any* social theory. As Popper says (1957): *'If it is possible for astronomy to predict eclipses, why should it not be possible for sociology to predict revolutions?'* Popper suggests that in vague and general terms it can, but that it cannot do so with precision because social events are so complex, so interconnected, and so qualitative in character that predictions of the kind which form the very basis of astronomy (that is mathematically based) simply are not possible in sociology. Not only that, but the predictions themselves *may* change the thing which is being predicted; an idea, he says, which is

a very old one. Oedipus ... killed his father whom he had never seen before; and this was the direct result of the prophecy which had caused his father to abandon him. This is why I suggest the name '*Oedipus effect*' for the influence of the prediciton upon the predicted event (or, more generally, for the influence of an item of information upon the situation to which the information refers), whether this influence tends to bring about the predicted event, or ... tends to prevent it.

Lucien Goldmann suggests some of the reasons for this (1969):

On the one hand, the historical and human sciences are not, like the physico-chemical sciences, the study of a collection of facts *external* to men or of a world *upon which* their action bears. On the contrary they are the study *of this action itself*, of its structure, of the aspirations which enliven it and the changes that it undergoes.

When it is a question of studying human life, *the process of scientific knowing, since it is itself a human, historical and social fact*, implies the *partial identity of the subject and the object* of knowledge. For this reason the problem of objectivity is posed quite otherwise in the human sciences than in physics or chemistry.

I have outlined some of the methodological difficulties in my book *Design in Architecture* of 1973. In reviewing the techniques available to psychology and sociology for measuring human attitudes, for instance, I suggested that:

They are open to faulty design, to errors in sampling, to a high rate of refusal in those invited to respond and even to bias on the parts of both the operator who sets up the scales and the interviewers who administer them. The respondents themselves may be ignorant or have highly personal reasons for not wanting to discuss it.

They may misunderstand the questions, find that none of the indicated responses matches their own attitudes, find that their attitudes change *because* the questions are being asked (effect of indeterminacy), give answers which they think the interviewer wants to hear or give tongue-in-cheek answers for the hell of it. There may be bias in the statistical analysis or final interpretation.

The crucial point here is that, unlike the planets which continued to move in their majestic courses *in spite* of the telescopes we have trained on them and the crazy hypotheses we have postulated to explain them, human beings are capable of detecting that they are being observed, however unobtrusive the observations; their behaviour in the laboratory will be different from their behaviour in 'real' life. And even if this were not true, if it were possible to write a perfectly objective *description* of human behaviour—with or without predictive powers—the writing down of that theory, its subsequent publication and dissemination would in *themselves* add to people's experience, and thus *may* change their behaviour. Given that designing itself is an act of human behaviour then—in spite of Simon's attempts to describe design as one of his *Sciences of the Artificial* (1969)—theories of design, in fact, prove to be even more elusive than more general theories of human behaviour. For even if one could devise an objective theory of design, the act of writing it down and publishing it would (hopefully) lead to its being read by designers, thus changing their behaviour, and so on. Indeed, if it did *not* change their behaviour, there was no point in one's writing, and their reading it in the first place. If it succeeded in changing the behaviour of designers it would, indeed, be more powerful than any 'real' theory. A real theory, by its very nature, does not and cannot change the behaviour of what has been observed. It is merely a description, with predictive powers, and that is its strength. Pseudo-theories have quite different strengths. They add to people's experience in ways which, hopefully, will change behaviour. In that sense they are prescriptive rather than explanatory and predictive.

To take only one of Popper's examples: Karl Marx's so-called theories include first of all a very tendentious history of society divided into a number of periods which suited his particular purposes, including:

(1) Man on his own (Robinson Crusoe) in symbiotic relationship with his envrionment, and therefore satisfied.
(2) Man living a primitive, communal life such as Marx thought (wrongly) appertained to the Indian village. (He quite forgot about the caste system.)
(3) Slavery in which the workers were *owned* by those who exploited them.
(4) Feudalism in which they were not owned but obliged to them by liege laws.
(5) Capitalism in which they were exploited by the bourgeoisie who employed them in factories.
(6) Communism in which, instead of being owned by private individuals, all means of production were to be owned communally, thus the imagined richness of Marx's Indian village life was recreated at a much higher economic level.

Within his tendentious history Marx described, quite accurately, the exploitation, the degradation, the alienation to which his proletarian workers were subject by being bound to the machine. He then presented a vision of what society could be like and it is that vision, above all, which has fired men's imaginations in so many parts of the world. Half the world now has some form of Marxist Government, although many would dispute whether Marx's vision actually has been realized in any of them. The point is that Marx's combination of tendentious history, description of present evils, and vision of what might be did change the course of human history, a thing which no true theory by definition (objective description with predictive powers) could possibly have done.

Curiously enough, the most influential design theorist of the twentieth century adopted similar strategies. Le Corbusier wrote his own tendentious history of architecture; he then described the evils of the late nineteenth century city and followed this by presenting his vision of what cities ought to be like. This too captured people's imaginations and Le Corbusier had the advantage over Marx in that in addition to describing his 'ideal' state of things he could go on to present actual built examples of houses, housing, and other building types of the kinds he had in mind. Others could then use these exemplars as bases for their own work.

The point obviously is that Marx and Le Corbusier changed the world because they wrote pseudo-theory and presented visions. So where does this leave the genuine theorist of design? Certain aspects of design obviously are susceptible to theory-building: at the scale of architecture this obviously is true of anything to do with materials—their properties and uses in building structure, construction, environmental control, and so on. A certain amount of true theory also may be possible concerning human physiological responses in building—to lighting, acoustic, thermal, and other environmental stimuli. But we are unlikely to build any true theories concerning human behaviour in buildings because of Popper's Oedipus effect, nor—for reasons we have already explored—can there be any true theories of design.

Are we then to be subject to the mere vagaries of pseudo-theory—to the seductive 'visions' of future Le Corbusiers? Yes, we are, but the seeds they plant can only germinate if they fall on fertile ground. Industrial society was—and still is in many places—as rotten as Marx said it was; that is why so many people still embrace his vision. The nineteenth-century city was as cramped, dark, and unhealthy as Le Corbusier said it was, which is why his particular visions have been built on such an enormous scale. Marx and Le Corbusier detected deficiencies in the world around them and made seductive proposals for their correction, which is why their ideas were adopted, why they were able to change the going paradigms. But their visions too had their faults. There is no Marxist state in the world in which people live in the freedom which Marx intended. With the single exception of Allende's Chile, every Marxist state so far has been imposed by revolution and sustained by the twin mechanisms of terror and bureaucracy. And Le Corbusier's city, as we very well know, is seen as bleak, hostile, and inhuman by those who have to inhabit it.

Of course, there are differences between Le Corbusier and Marx which can be clarified with reference to Kuhn's modified paradigm. Taking the components in turn: Marx's paradigm did not, and could not, contain any technical knowledge, laws of nature, and so on: these are not in the nature of human interaction. They certainly included professional skills and ways of modelling the world, and Marx also presented exemplars of ideal societies, although these were very tendentious indeed. The real weight of Marx's pseudo-theories, of course, lay in their ideological content.

Le Corbusier's paradigm presupposed certain professional skills and ways of modelling the world; they also contained exemplars (Le Corbusier's own buildings) and contained their own (*Brave New World*) ideology. But unlike politics, building obviously is bound by the laws of nature. It was Le Corbusier's ignorance, and misunderstanding of these laws which lead to so many difficulties in the performance of buildings within his paradigm. That is why in correcting their faults, we *do* need that access from research which Meyer, Maldonado, Martin, Llewellyn-Davies, and others thought we should have.

So, one could almost begin to plot a range of activities in research, theory-building, and design and even to show that these are the province of certain professions. One possibility is shown in Figure 1.

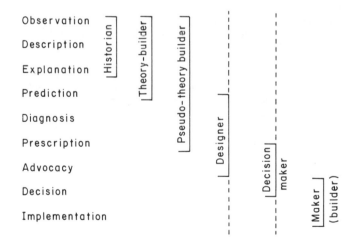

Figure 1

My own view, which, of course, will be controversial, is that these activities increase in difficulty in ascending order from observation at least to decision: that whilst explanation, of the scientific kind, certainly requires an access of creativity in terms of hunch, or conjecture in Karl Popper's sense, prescription, advocacy, and decision require at least as much creativity, while the latter two also require an understanding of human interaction, and a charisma which are simply not necessary for the scientist in his lab.

So, far from regretting the fact that the scientist may believe that he alone is capable of true research and 'real' theory-building, I am quite happy to leave these with him, on the grounds that prescription, advocacy, and so on are much more difficult things to do, and much more rewarding.

REFERENCES

Broadbent, G. (1973), *Design in Architecture*, Wiley, Chichester.

Copernicus (1510). See Kuhn T. S. (1957), *The Copernicus Revolution: Planetary Astronomy in the Development of Western Thought*, Harvard University Press, Cambridge, Mass.

Galileo, G. (1632), *Dialogue on the Great World Systems* (1953), University of Chicago Press, Chicago.

Goldmann, L. (1969), *The Human Sciences and Philosophy* (trans. White, H. L. and Anchor, R.), Cape, London.

Gropius, W. (1956), *The Scope of Total Architecture*, Allen and Unwin, London.

Kuhn, T. (1962), *The Structure of Scientific Revolutions*, University of Chicago Press, Chicago.

Kuhn, T. (1970a), 'Reflections on my critics', in Lakatos, I., and Musgrave, A., *Criticism and the Growth of Knowledge*, Cambridge University Press, London.

Kuhn, T. (1970b), Introduction to *The Structure of Scientific Revolutions* revised edition, University of Chicago Press, Chicago.

Lakatos, I., and Musgrave, A. (1970), *Criticism and the Growth of Knowledge*, Cambridge University Press, Cambridge.

Laplace, P. S. de. (1812), 'Theorie analytique des probabilities'; translated as 'Concerning probability', in Newman, J. R. (ed.) *The World of Mathematics*, vol. 11, pp. 1325–33, Allen and Unwin, London (1960).

Le Corbusier (1923), *Vers une Architecture*, Cres, Paris.

Llewellyn-Davies, R. (1961), *The Education of an Architect*, H. K. Lewis and Co. Ltd, London; also in *Architects Journal*, 17 November 1960.

Masterman, M. (1970), 'The nature of paradigm', in Lakatos, I., and Musgrave, A., *Criticism and the Growth of Knowledge*, Cambridge University Press, London.

Marx, K. (1866–67), *Capital*, vol. 1 (trans. Fowkes, B., 1976), Penguin, Harmondsworth.

Popper, K. R. (1957), *The Poverty of Historicism*, Routledge, London.

Popper, K. R. (1963), *Conjectures and Refutations*, Routledge, London.

Ptolemy, Peters, CHF, and Knobel, E. B. (1915), *Ptolemy's Catalogues of the Stars: A Revision of the Almagest*, Carnegie Institution, Washington.

Schnaidt, C. (1965), *Hannes Meyer: Buildings, Projects and Writings* (English version by Stephenson, D. U.) Tiranti, London.

Simon, H. (1969), *The Sciences of the Artificial*, MIT Press, Cambridge, Mass.

4.4 Design Creativity and the Understanding of Objects

Janet Daley

There has probably been no question more urgent or controversial in the brief history of design theory than that of the
status of knowledge claims within the discipline itself. Perhaps
the crucial questions could be characterized most helpfully as:
are the processes by which designers make their decisions
susceptible to systematic examination? If so, what sort of
examination? Are such processes 'conscious' in all their facets
and if, as seems clearly to be the case, they are not, what are the
consequences of attempting to translate them into the terms, or
the currency, of that which is examinable in the straightforward
sense? What, precisely, is the nature of the knowledge which
designers *have* and *use* (although this may be a false distinction)
when they engage in practice and, further more—a quite
different question—what is the nature of the metaknowledge of
such skills and practices to which design theorists aspire?

It seems beyond doubt that the answers to such questions, if
indeed there are answers, can only be formulated in epistemological terms. Epistemology has, since Descartes, taken as one
of its chief concerns, the status of knowledge claims. As well as
contributing to this meta-issue of design theory, that is, to the
issue of theoretical methodology, epistemology has a much
more fundamental role to play, in establishing an account of
how it is that we understand our relationships with objects.
Which is to say, how it is that we manipulate our conception of
reality in such a way as to make innovations in spatial relations
and, at times, create wholly new object configurations. In fact,

Originally published in *Design Studies*, **3** (3) (1982), 133–7. Reproduced by
permission of Butterworth and Company (Publishers) Ltd.

I do not see what I have called the more fundamental issue of design theory—that of our understanding of objects—as being separable from the meta-issue of knowledge claims about design process. If I am right in my tentative formulation of the former, then some conclusions follow inescapably about the latter.

CLASSICAL EPISTEMOLOGY

To appreciate the role which epistemology as a discipline must play in any ultimate account of creativity in terms of objects and their relations, it is necessary to have something of a grasp of the history of this concern in epistemological literature. Although the most familiar modern epistemological accounts of the problem of objects stem from Cartesian rationalism, a recognition of the dilemma goes back to the pre-Socratics, and may be seen clearly in Plato's attempts to account for universals in terms of a metaphysical alternative reality.

The paradox of a *perceived* world of material objects having the status of reality has bedevilled western philosophy from its beginnings: since our perceptions of objects seem to be logically separable from any objective existence that the objects may have and, further, since any knowledge that we have may be seen as only some sort of compound of our perceptions organized, perhaps, on the basis of *a priori* principles, how can we be justified in positing the actual objective existence of the perceived world, let alone in making any verifiable knowledge claims about it?

Descartes, it was fashionable to claim until recently, locked us unequivocally into the dilemma with his mind–body dichotomy: consciousness, equipped with certain innate ideas, rendered external reality intelligible by applying those *a priori* concepts which were a function of pure rationality. It was only the conception of an innate idea of 'objectness' which could account for our ability to see the world as consisting of continuous material objects, separable from their changing and fluid qualities which were a function of sense perception.

Pure reason inhabiting Mind, and sense experience being a function of Body, offered a conceptual framework in which the criteria for knowledge, and for certainty, were describable. Knowledge was a product of reason; sense experience was, as often as not, a source of confusion. Cartesian rationalism was clearly descended from Platonic Idealism in its distrust and disregard for the evidence of sense experience and its conception of true knowledge as being in another realm (logically, rather than metaphysically for Descartes) from perceptual appearances.

For Descartes, all that we may *perceive* are qualities: yellowness, hardness, roundness, etc. That those qualities *inhere* in (in Descartes' own famous example) a particular piece of wax, is a notion which no amount of perceiving could give me. Furthermore, when I place this piece of wax by the fire so that all of its perceptible qualities change—it is now soft, liquid, white and occupying a larger amount of space—I am still prepared to say that it is the same piece of wax as I had earlier perceived. This continuity or integrity of objects through a perceptual change, and indeed, through time and space, could only be imposed by an intellectual act which was independent of perception.

The empiricists rejected Descartes' innate ideas on ideological grounds: infatuated with empirical science, they wished to establish principles of knowledge which were based entirely on empirical foundations. Having discounted the possibility of conceptual schemata preceding sense experience, they were forced to fall back on hidden principles of organization by which simple ideas (of perceptible qualities) might be organized into the complex ideas of objects as they were experienced. They were faced, in other words, with the problem of delineating the conceptual mechanism whereby our sense experiences which are particular and transitory, give rise, in and of themselves, to an intelligible construction of the world as universal and permanent.

Empiricism

George Berkeley, probably the most undervalued figure in British philosophical history, saw clearly the logical flaws in Locke's attempts to maintain an objective reality while insisting that sense experience was the only source of knowledge. If there was such a reality then it was, by definition, outside of our experience and its existence could not be posited. Hence, to be *was* to be perceived, and all that can be said to exist are perceptions. Berkeley saw correctly that what he had created in forcing empiricism to be consistent was an unintelligible universe in which there were no identifiable objects, no continuous features, no recognizable—and hence, knowable—phenomena. No predication was possible since there were no permanent objects of which to predicate any qualities. No prediction was possible since the 'world' consisted of nothing but transitory, unidentifiable, particular sense impressions.

Clearly such a view was unacceptable and Berkeley, to his eternal disrepute in the history of philosophy, rescued the intelligibility of the universe by recourse to theology. It is the omniscient, all-perceiving mind of God which guarantees the continuous existence of the world. Because God perceives everything all the time, we are not faced with the alarming prospect of the world popping in and out of existence every time we shut our eyes. Students often get the impression at about this point that the most significant questions of western epistemology are of the order: how can you know whether the light stays on in the fridge after you have shut the door? Certainly, the issues of doubt and reality take on a farcical quality especially as they descend into solipsism with David Hume. But one can only attempt to make clear the sense in which it is something of an epistemological miracle that a world of objects with an independent existence, is a fact of the nature of our experience.

Hume, having carried scepticism to its ultimate logical conclusion, undermined belief in any phenomenon or relation which could not be directly experienced, and thereby eliminated from credibility induction (and hence, scientific prediction), memory, the external world, and the very notion of self as a prevailing substantive. Having elevated direct experience to an absolute criterion of knowledge, Hume had effectively dismantled the concept of Mind. Descartes' 'I think, therefore I am' had become, 'I think, therefore thought is'.

Perception and *a priori* concepts

Something of a synthesis of rationalism and empiricism which, however, transcended many of the limitations of both, is found in Kant. For Kant, the categories of *a priori* concepts and knowledge-from-experience were not mutually exclusive. As in all of the greatest philosophical solutions, Kant sought his answer by altering one of the most fundamental presuppositions of the enquiry. He reconstructed the very notion of perception itself, seeing it not as a passive process with Mind as pure receptor, but an active process in which the mind is an agent. We do not so much *perceive* the world (in a passive sense) as *construe* it and the process by which this construction is achieved is not a mechanistic combining of sense data, or even an application of given innate ideas. The *forms* of our perception are determined by the limits of our *a priori* conceptual framework.

Thus, our experience itself is sieved, as it were, through the preshaped slots of those conceptual categories which are all that is given to us. Without sense experience there would be no ideas—but without the *a priori* conceptual categories with which we come equipped, there would be no intelligible experience. We are neither empty receptacles on the empiricist model, nor self-contained vessels of inborn truths on the Cartesian one. But above all, the very concept of perception is meaningless without a logically prior system of categorical organization. Kant says, in the *Prolegomena*, 'The understanding does not derive its laws from, but prescribes them, to nature.'

PERSISTENCE OF OBJECTS

The limits of our understanding, for Kant, were based fundamentally on our ability to construe a reality of continuous *objects*. It was the crucial concept of what we would now call object constancy, which Kant took to be indispensable to a world which was to be susceptible to any sort of description, which was comprehensible within the logic of our discourse. It is only through the concept of objects which persist through time and space that we can have any notion of change, causality or identity. Predication and prediction are comprehensible only in a world which consists of identifiable particulars.

Thus, the logic of such identification and the criteria for recognition of object identity are central concerns for Kant, as indeed they have been for epistemology ever since. These two notions seem indispensable: that the cognitive structuring of a world of objects is central to human knowledge and understanding, and that such a structuring is an *act* of mind (though obviously not a conscious one) and not a function of the world, which impinges on mind. Kant felt that the limits of our understanding, although they constituted the bounds of logical possibility, were themselves contingent. That is, the limits of what is, for us, logically possible, are determined by our natures.

We construe the world in the way that we do, because we are the sort of creatures that we are. If we all had eyes like electron microscopes we would have no conception of fixed objects but only of molecules in a constant state of flux. Our sensory apparatus, linked to a particular rate of perceptual processing, freezes phenomena into a state in which the concept of objectness becomes the basic reference. But there is nothing logically necessary about these particular conceptual limitations as modern physics has made abundantly clear to us. Objects as such, are a function of our experience, of our way of experiencing.

The consequence of this for our understanding of design, is a realization that an imaginative manipulation of objects in space and time is a condition of all intelligible human experience, and if we are to understand the rather special manipulations which designers perform on the outer frontiers of ordinary understanding, then we must see it within this context—within an understanding of the fact that 'imagination', in the eighteenth-century sense, is fundamental to all experience of the world, and that the most mundane seeing of an understandable world is, in a very real sense, a creative act.

It is intriguing that we can find contemporary descendants of Kant, not only in philosophy proper but in disciplines which, a generation ago, seemed to be hived off quite irrevocably from their origins in seventeenth-century natural philosophy. Developmental psychology and linguistics both seem now to be engaged in examining issues which bear directly on the classical epistemological controversies surrounding *a priori* perceptual schemata and innate conceptual structures. There is, for those whose tastes are so inclined, a pleasing regeneration of holistic theories of mind which seems to be reversing the historical trend toward fragmentation both of disciplines and of examinable phenomena. There is no longer quite such a clear-cut distinction between a psychological and a philosophical issue.

INNATENESS OF CONCEPTS

In the work of Dr Tom Bower of Edinburgh University, UK for example, there are inevitable implications for epistemology. It would seem from Dr Bower's evidence that neonates— infants too young to have had any tactile experience of the manipulation of objects—'know', in some sense of that word, that objects have solidity and extension (that they occupy a three-dimensional space) and have quite sound expectations about the behaviour of objects. An object which is apparently growing larger is perceived by a neonate as coming closer; an apparent 'object', created by visual illusion, is grasped at and evokes alarm with its non-existence; an object approaching the infant's head on a hit path evokes alarm, where one approaching on a miss path does not.

Most interesting of all, perhaps, is that infants only weeks old manifest expectations about the continuity and integrity of line, about geometric shapes. A triangle intersected by a bar which obliterates its middle section seems to be understood as continuous and not as two separate items (see Figure 1). Bower's experimental work, in an uncanny way, evokes an echo of Descartes himself: there would seem to be, at the most minimal interpretation, an innate predisposition to perceive three-dimensional objects existing in a continuous space and

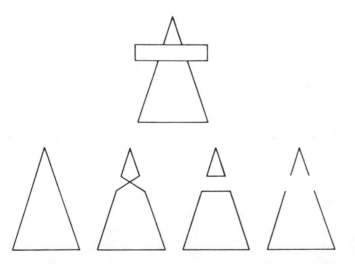

Figure 1
Eight-week-old babies used the rule of 'good continuation'. Trained to respond to the triangle intersected by the bar, they unhesitatingly transferred all their responses to the complete triangle rather than any of the other three alternatives in the figure.

time. Dr Bower makes the point in vocabulary which is akin to discussion of purely epistemological issues:

Jane Dunkeld compared babies' responses to an approaching object with their responses to an approaching hole. In terms of contour expansion the two are identical, and yet it was only the approaching object that elicited defensive responses. An object of course, is perceptually different from an aperture. Newborns seem able to pick up these differences and *seem to realize* that one perception signifies emptiness and the other hardness and solidity. (Bower, 1977, emphasis added.)

It is worth recalling that Descartes felt that the primary qualities which must be present in an object (for it to count as existing) were extension (the taking-up of space), solidity and figure (the having of a shape) and that these notions, implicit in objectness, must be innate because they could not be learned.

Other researches in developmental psychology (Ahrens, Spitz, Wolf, Fanz, Trevarthan) have concentrated on establishing innate schemata for sociality. That the human infant seems to be equipped to initiate and respond to social interaction and even to recognize, in a schematic way, fellow human beings, has been established in over 20 years of very interesting experimental work. That there may be a close connection between such an innate apparatus for social behaviour and the cognitive structures suggested by Dr Bower's works is an argument to which I shall return.

Innateness of language

Probably the most well known version of the innateness hypothesis is that contained in Professor Chomsky's linguistics. It is impossible, within the limitations of this paper, to discuss Chomsky's theories in any detail. His ideas are the subject of an examination in a paper by my colleague, Noel Sharman (1984), and therefore it only remains for me to comment on Chromsky's role in epistemological debate.

What is of most concern for our purposes is Chomsky's account of creativity which he describes as being based on a process of generative first principles. It is debatable (and, indeed, is debated) whether Chomsky is adopting a hard line, which would read something like:

There are certain universal innate schemata which provide the essential structures by which all grammars are generated and it is the possession of those structures which makes possible infinite innovation within the acquired specific language.

An alternative softer line which many commentators feel to be implicit in Chomsky's later writing, could be paraphrased:

There are universal innate predispositions toward structural grammars which have a certain kind of logical consistency. What is given *a priori*, are principles of organization which make linguistic innovation possible.

CREATIVITY AND INNATENESS

On either interpretation, what is of interest to us here is the argument for creativity in terms of innate generative principles, and the implicit rejection of the notion that any adequate account of creativity could be based on a simple accumulation of sensory input. In looking for a philosophical model for both Chomsky's theory of language innateness and the developmental psychologist's hypothesis of innate schemata for sociality, it seems to me that the later Wittgenstein provides a peculiarly congenial framework. Wittgenstein conceived of language as rule following behaviour of a societal nature, indeed, he described the criteria for the meaningfulness of language as being necessarily public and only understandable as a funcion of a particular 'form of life'.

In other words, it is only man's life in a particular social context which allows for the possibility of intelligible language. The principles of organization which would enable us to have the *a priori* understanding of rule-following necessary in Wittgenstein's formulation, could easily be seen in terms of Chomsky's hypothesis. Similarly, the systematic impulses toward social interchange and interpersonal communication which developmental psychology documents, also seem compatible with the Wittgensteinian argument for the precedence of social communication in any account of how language means.

It will be obvious at this point, that what I am hinting at is nothing less than an embryonic theory of mind. Let us accept, for the sake of argument, that the predispositions toward, and even the schemata for, social intercourse and object concepts are innate and that, following Wittgenstein, it is the forms of social life which provide the basis for the meaning of language, and further, that it is the schema for object constancy which makes possible an intelligible sense experience. What might follow from this? The sense which we impose on our perceived experience, which centres on a world of continuous objects, and the value structures arising from sociality, are what makes knowledge possible.

One might suggest a hierarchy of these epistemological levels. First, and most fundamentally, the biological imperatives of the human being give him an absolute dependence on sociality. These biological needs have as their consequence innate schemata for social response and intercourse, and these impulses toward social interdependence underlie psychological predispositions toward communal behaviour.

It is from these psychological inclinations that value presuppositions arise. I mean by 'value', not only moral consciousness in the usual sense, although certainly that is included, but also in the sense of the ordering of conceptual priorities. That is, without value structures on the most fundamental level, we

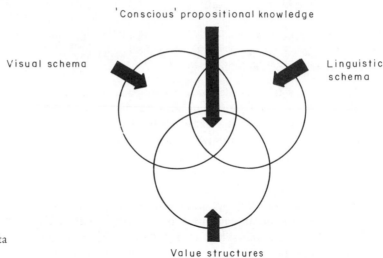

'Conscious' propositional knowledge

Visual schema

Linguistic schema

Value structures

Figure 2
Overlaps of the various schemata
constructed by the mind.

should be unable to make even the most rudimentary sense of our experience. Only by having criteria of judgment for what counts as an event, what constitutes a phenomenon, could we ever proceed to construct any systematic knowledge out of a vast myriad of sensory inputs, most of which are literally beneath our notice (because they do not constitute any significant part of the communal picture of the world which is a function of our social life).

The most superficial (and evolutionarily recent) level of this hierarchy, I suggest, is propositional knowledge, the concept of which is only possible as a consequence of the value structures I have described. Only after the fact of those *a priori* judgments of value which underlie our conceptualization of experiences, can we conceive of the notion of an empirical fact.

Knowledge and innate schemata

Perhaps my most controversial claim would be that all of the propositional knowledge (and consequent reasoning) which we take to be the content of our intellectual activity is but the small intersection of a set of systems. That is, the mind may not have *a* systematic way of knowing or conceiving, the schemata of which can be definitively described, but it may have a number of innate capacities for constructing schemata, the logics of which may only minimally overlap, i.e. be translatable into the terms of the others. Thus, conscious mental activity, with its language-based emphasis on propositional knowledge may be the area in which these various systems intersect with that of verbal discourse (see Figure 2).

CONCLUSIONS

What is the import of this for theories about how designers manipulate ideas of objects? If my fledgling theory of mind is correct then several things follow:

(1) The capacity to visualize an as-yet-unmade object and to manipulate spatial relations lies at a fundamental level of cognitive ability, and any explanation of this capacity, or the processes which it involves, must address itself to those *a priori* structures which make any conceptual construction of a world of objects possible.

(2) Only a relatively small (and perhaps insignificant) area of that system of knowing and conceiving which makes designing possible may be amenable to verbal description. To talk of propositional knowledge in this area, or to make knowledge claims about the thinking processes of designers, may be fundamentally wrong-headed. The way designers work may be inexplicable, not for some romantic or mystical reason, but simply because these processes lie outside the bounds of verbal discourse: they are literally indescribable in linguistic terms.

(3) This may go some way to explain the attraction that artistic and design creativity have for us: their processes transcend the bounds of verbal discourse. They make leaps, as it were, into primeval levels of mental life, manipulating the very constituents of our picture of physical reality. When artists and designers create, they are delving into the very order of the universe in perceptual terms.

But if we can never, in any describable sense, know what happens when artists and designers practise, what exactly can we say on the subject? For one thing, if my arguments are valid, then it is a fatal mistake to regard design processes as straightforwardly rational, unconnected to man's social nature and his consequent value structures. We must portray the designer not as an intellect executing decisions but as a human being whose entire mental life is immersed in the parameters and priorities of his existence as a social being.

Thus, it is not only for humane or conscientious reasons that I would argue that man's social nature, and his consequent moral consciousness, must be integrated into any theoretical model for a designer's activities, but because I believe that these levels of experience lie at the very basis of the activity itself.

But if there is so little that we can describe of the activity, what sort of knowledge is it that the designer can be said to possess? Not, obviously, a compendium of propositional knowledge. If design and artistic creativity are experimentation with our perceptual limits, then they are irrevocably outside the realm of verbal description. We cannot describe the bounds of our experience because, as both Kant and Wittgenstein made clear, we can never step outside them. But this is the true province of design, which is not a meta-language nor a set of deductions, but a systematization of our experience of the physical world. The ways in which we symbolize and represent that kind of systematization may be the most important clues to how we make sense of the world. Only by remaining true to the nature of the level of experience which is entailed can we hope to reach any real understanding of the phenomena for which we are trying to account. These boundaries may be, in effect, where philosophy ends and art begins.

Acknowledgments

I am indebted to my student, Gary Johnstone, for the references to Trevarthan.

REFERENCES

Berkeley, G. (1915), *A Treatise Concerning the Principles of Human Knowledge*, Open Court Publishing Company, La Salle, IL, USA.

Bower, T. (1977), *The Perceptual World of the Child*, Fontana/Open Books, London.

Bower, T. (1979), *Human Development*, W.H. Freeman, San Francisco, CA, USA.

Chomsky, N. (1957), *Syntactic Structures*, Mouton, The Hague, The Netherlands.

Chomsky, N. (1966), *Cartesian Linguistics: a chapter in the history of rationalist thought*, Harper and Row, London.

Chomsky, N. (1968), *Language and Mind*, Harcourt, Brace and World, New York, NY, USA.

Descartes, R. (1912), *A Discourse on Method and Selected Writings* (translated by John Veitch) J.M. Dent and Sons, London.

Fanz, R. L. (1964), 'Visual experience in infants: decreased attention to familiar patterns relative to novel ones', *Science*, **146**, 668–70.

Freedman, D. (1964), 'Smiling in blind infants and the issue of innate vs. acquired', *J. Child Psychol. Psychiatry*, **5**, 171–84.

Hume, D. (1894), *An Enquiry Concerning the Human Understanding and an Enquiry Concerning the Principles of Morals* (edited by L.A. Selby-Bigge), The Clarendon Press, Oxford, UK (second edition 1936)

Hume, D. (1896), *A Treatise of Human Nature* (edited by L.A. Selby-Bigge), The Clarendon Press, Oxford, UK.

Kant, I. (1929), *Critique of Pure Reason* (translated by Norman Kemp Smith), Macmillan and Co, London.

Kant, I. (1950), *Prolegomena to any Future Metaphysic* (translated by Lewis Beck White) Liberal Arts Press, Incorporated, New York, NY, USA.

Kenny, A. (1973), *Wittgenstein*, Penguin, London.

Locke, J. (1894), *An Essay Concerning Human Understanding* (edited by Alexander Campbell Fraser), The Clarendon Press, Oxford, UK.

Sharman, N. (1984), 'Transformational theory: A linguistic paradigm for the ability to innovate in object perception' in Langdon, R. et al. (eds), *Design Policy*, The Design Council, London.

Spitz, R.A. and Wolf, K.M. (1946), 'The smiling response: a contribution to the ontogenesis of social relations', *Genetic Psychol. Monographs* **34**, 57–125.

Trevarthan, C. (1977), 'Descriptive analyses of infant communication behaviour', in Schaffer, H.R. (ed.) *The Mother–Infant Interaction*, Fontana, London.

Trevarthan, C., and Grant, F. (1979), 'Infant play and the creation of culture', *New Scientist*, pp. 566–9.

Trevarthan, C., and Houbley, P. (1978), 'Secondary intersubjectivity, confidence, confiding and acts of meaning in the first year', in Locke, A. (ed.) *Action, Gesture and Symbol*, Academic Press, London.

Wittgenstein, L. (1963), *Philosophical Investigations* (translated by G.E.M. Anscombe), Basil Blackwell, Oxford, UK.

Wolff, P.H. (1963), 'Observations on the early development of smiling' in Foss, B.M. (ed.) *Determinants of Infant Behaviour*, Vol. 2, Methuen, London.

Part Five

The History of Design Methodology

Introduction

The first four Parts of this book have recorded some of the main features of the development of design methodology over a period of twenty years, from the early 1960s to the early 1980s. In that period the study of design methodology was pursued by a loosely connected international network of scholars and researchers (and an occasional design practitioner) representing what has often been regarded as 'the design methods movement'. As well as making their contributions to the development of design methodology, some members of this 'movement' have also commented from time to time on the progress of the 'movement' in general, and on their own changing attitudes towards design methods. In this final Part some of the more interesting and significant of these comments have been collected together to provide a more informal version of the history of design methodology.

The movement, if that is what it has been, was kept alive—like other minority social movements—by conferences and small publications. For many years the principal publication in the field was a newsletter published in the USA by the Design Methods Group—the *DMG Newsletter*. In 1971 the *DMG Newsletter* initiated a series, based on a standard questionnaire to contributors, on 'The state of the art in design methods'. Two of the responses (in the form of interviews), by Alexander and Rittel, are reprinted here, offering rather different views of design methodology as it stood at the beginning of the 1970s.

Although Christopher Alexander's early work on decomposition techniques (see Chapter 1.2) had been a powerful and original stimulus to the initial emergence of a design methods movement, by the time of his DMG interview in 1971 he had apparently become totally disillusioned with design methods and especially with design methodology. His advice is 'forget it; forget the whole thing', because he feels that the development and study of design methods has failed to contribute to better design.

Alexander claims that his original motive for getting involved with design methods was because he wanted to design beautiful buildings—buildings as good as traditional, vernacular architecture. Influenced also by his background in mathematics, he wanted to be able to describe a logical, step-by-step process for design because that would ensure that he really knew what he was doing: 'the definition of a process, or a method, was just a way of being precise, a way of being sure I wasn't just waffling'. However, in applying his early work he found that he was able to go fairly directly to his schematic design diagrams without the exhaustive prior analysis of problem requirements. This realization led him into his work on 'patterns'.

In the interview, Alexander is frequently scathing in his remarks on design methods and methodology, suggesting that the methods had become irrelevant to real design and that the study of methodology had become a pointless preoccupation and excuse for people with a fear of engaging in real design activity. It was bold of the *DMG Newsletter* to publish the interview, since it seemed as though the *Newsletter* was publishing its own death warrant. There was certainly a feeling at the time that the design methods movement was dying. But a major attempt at resuscitation was made a year later, with the publication of the interview with Horst Rittel, another leading figure in the design methods movement.

Rittel accepts that much of the early design methods work seems to have been fruitless, and proposes a concept of 'generations' of design methods. This was a brilliant idea because it allowed the old, or 'first-generation' methods to die a respectable death, whilst the movement itself could survive in the development of 'second-generation' methods. Understandably, the *DMG Newsletter* editors published this interview as a special *DMG 5th Anniversary Report*. The infant movement was alive and well!

According to Rittel, the first-generation methods were based on the wrong premises to be really useful in design. They had been drawn from the systems engineering techniques of military and space missions, and therefore were not wholly adequate to the 'wicked' problems of planning and design. In trying to apply these methods some lessons had been learned, and a new way forward could be discerned. Rittel proposes a number of principles for second-generation methods. Perhaps the most important of these is that the design process is based on an 'argumentative' structure, and that expertise and relevant knowledge is assumed to be distributed amongst a wide range of participants. Thus second-generation methods are intended especially for a more participatory approach, in which the role of the planner or designer 'is that of a midwife or teacher rather than the role of one who plans for others'.

This new emphasis was a reflection of the extensive moves in the late 1960s and early 1970s towards opening up the processes of planning and design so as to include the participation of laypeople. Both Rittel and Alexander suggest greater participation in design as a key reason for their changed perspectives on design methods.

Another leading figure in the design methods movement whose perspective changed quite radically in the early 1970s was J. Christopher Jones. Like Alexander and Rittel, Jones rejected much of the early work in design methods, but seemed prepared to go much further in a search for new approaches not only to planning and designing but also to living in general. In a covering letter accompanying the publication of the paper reprinted here, Jones explained that it is 'more a poem than an article', and that 'What it says, among other things, is that a new flexibility is wanted now in designing as in living.' In the search for this new flexibility Jones turned to the arts and particularly to works (such as those of the musician John Cage) which relied on chance or random processes for their composition. He began composing essays (about design, planning, technology, and life) which incorporated several sources in a randomized format. This is how he explained the composition of the essay on 'How my thoughts about design methods have changed during the years':

I wrote it quickly one Saturday morning (in October 1974) in response to Day Ding's request that I give a lecture at Illinois about how my ideas of design have changed. Earlier I'd tried out his suggestion in an informal talk to some architecture students at VPI (Virginia Polytechnic Institute). One of the students taped the talk so I composed the paper for Day by mixing extracts from the tape with my rough notes from which it was given. After each of my notes, IN CAPITALS, is whatever came on the tape at randomly chosen points. Not exactly random: I transcribed whatever came immediately after the numbers 000 100 200 etc., on the position indicator. The line lengths on the page, for the tape-extracts, are chosen partly by breath pauses and partly by 'units of thought'. The first two remarks were by someone else.

The essay traces Jones' involvement in design from his early work in ergonomics in the 1950s, through his work in the mainstream of design methods in the 1960s, to his experimenting with chance processes in the early 1970s. What emerges is a continuing concern with trying to resolve the apparent conflicts between rationality and intuition, logic and imagination, order and chance. Jones reacted strongly against the direction he perceived the design methods movement taking in the early 1970s, against 'the machine language, the behaviourism, the continual attempt to fix the whole of life into a logical framework'. His new contributions to the movement have been difficult for many to accept; after all, chance, or accident, is usually regarded as the antithesis of design. However, the use of randomization procedures is simply a technique which (like some other design methods) embodies a rational decision to let chance play a major role in the process of composition.

Like Alexander and Rittel, Jones was anti-expert—against the planners who decided how everyone else should live—and this view predominated through the development of second-generation design methods in the mid-1970s. By the end of the 1970s it was time for someone to suggest the need for a third generation. In his paper on 'The development of design methods' Geoffrey Broadbent chronicled the rise and fall of both first- and second-generation design methods, and went on to propose the premises of the third generation.

Other commentators have frequently found it difficult to cite major examples of the use of design methods. But Broadbent cites examples of the 'successful' application of first- and second-generation methods, and uses them to show just what was wrong with those methods. His example of successful first-generation methods is Disney World in Florida. This planned 'community' uses rational techniques such as queuing theory to control the programmed activities of its temporary inhabitants, and this Brave New (Disney) World therefore epitomizes the fears of bureaucratic, behaviouristic planning expressed by Jones and others. Broadbent similarly undermines the claims of second-generation design methods by pointing to the straightforward architectural inadequacies of the buildings of the University of Louvain, Belgium, which were designed by participatory methods.

Broadbent offers a model for third-generation design methods based on Popper's 'conjectures and refutations' model of scientific method. He suggests that the role of the designer is to make expert design conjectures, but that these must be open to refutation and rejection by the people for whom they are made. This suggestion is meant to draw upon, and to synthesize, the better aspects of both first- and second-generation methods, and so—far from being unhappy about the development of design methods—Broadbent is able to conclude that 'design methods are alive and well'.

By the end of the 1970s it became clear that design methodology had passed through its crisis of confidence. Bruce Archer's question, 'Whatever became of design methodology?' was largely rhetorical. After allowing himself a brief glance backwards so that he can affirm that for him, too, the interest in design methods had been to create better designs, he takes a very positive attitude towards the current state of the art. The basis of this attitude is his belief 'that there exists a designerly way of thinking and communicating that is both different from scientific and scholarly ways of thinking and communicating, and as powerful as scientific and scholarly methods of enquiry when applied to its own kinds of problems'.

So for Archer, what is wrong with some of the mathematical and logical design methods is that they are 'the product of an alien mode of reasoning'. Designerly ways of thinking are quite different—but quite appropriate to the kinds of ill-defined problems designers tackle. Ill-defined, untamed problems are real problems of everyday life, and so the methods for tackling these problems are deeply rooted in human nature, Archer claims. This means that design methodology must be based on the study of fundamental, innate human capacities. There is a newly confident view, expressed by Archer, that design methods must not try to ape the methods of the sciences or the humanities but must be based on the ways of thinking and acting that are natural to design. It is from this viewpoint that design methodology can be seen to have a valid role to play in the development of design research, design education, and design practice. Design methodology is, indeed, alive and well.

FURTHER READING

Both Alexander and Jones have provided additional comments on their changing views in the new introductions to the paperback editions of their early principal works: Alexander, C. (1971), *Notes on the Synthesis of Form*, Harvard University Press, Cambridge, Mass.; Jones, J.C. (1980), *Design Methods: seeds of human futures*, Wiley, Chichester.

Early review papers of the design methods field were Jones, J.C. (1966), 'Design methods reviewed', in Gregory, S.A. (ed), *The Design Method*, Butterworths, London; and Broadbent, G. (1968), 'A plain man's guide to systematic design methods', *RIBA Journal* (May), pp. 223–7. A more recent review is Cross, N. (1980), 'The recent history of post-industrial design methods', in Hamilton, N. (ed), *Design and Industry: the effects of industrialisation and technical change on design*, Design Council, London.

There are two principal journals by means of which one may keep up to date with design methodology: *Design Studies,* published by Butterworth Scientific Press, PO Box 63, Guildford, Surrey, UK; and *Design Methods and Theories*, published by the Design Methods Group, PO Box 5, San Luis Obispo, California, USA.

5.1 The State of the Art in Design Methods

Christopher Alexander
(Interviewed by
Max Jacobson)

Jacobson: What do you see design methodology as trying to do?

Alexander: Interesting question. Obviously the intent is to try and create well-defined procedures which will enable people to design better buildings. The odd thing is that in the vast proportion of the literature people have lost sight completely of this objective. For instance, the people who are messing around with computers have obviously become interested in some kind of toy. They have very definitely lost the motivation for making better buildings. I feel that a terrific part of it has become an intellectual game, and it's largely for that reason that I've disassociated myself from the field. I resigned from the Board of Editors of the *DMG Newsletter* because I felt that the purposes which the magazine represents are not really valuable and I don't want to be identified with them. And there is so little in what is called 'design methods' that has anything useful to say about how to design buildings that I never even read the literature any more. There's an amazing gap between the avowed intent and the actual intent of the field. If the intent of the field actually had to do with making better buildings and better cities, I could get interested in it but as it is, I'm not. I no longer see my own work as part of it at all because I definitely *am* concerned with trying to make better buildings. That remark of Poincaré is very much to the point: 'Sociologists study sociological methods; physicists study physics'. The idea

Originally published in *DMG Newsletter*, **5** (3) (1971), 3–7. Reproduced by permission of The Design Methods Group.

that you can study methods without doing and studying design is a completely mad idea as far as I'm concerned.

Jacobson: What about the work on developing techniques to *evaluate* designs and buildings?

Alexander: I'm very suspicious of that also. I know it's a commonly accepted idea in this culture that the critics of something do not necessarily have to be artists themselves. This is commonly accepted in music, literature, painting. I don't agree with that at all. I think it's absurd. I don't think one can criticize things valuably unless one is at least attempting oneself to make things of whatever sort are being discussed. I find the critics of architecture not helpful. If a group of people, forming a subdiscipline with the avowed intent of separating themselves from the practice of design, attempt to evaluate buildings, they are not going to shed much useful light on the subject.

Jacobson: What about those techniques being developed which are specifically intended to aid the act of design, such as brainstorming?

Alexander: Brainstorming—I find it incredibly naïve and odd to treat that as a subject of study in itself. I feel that that kind of self-consciousness about one's activities actually removes one from the spirit of the matter. There was a conference which I was invited to a few months ago where computer graphics was being discussed as one item, and I was arguing very strongly against computer graphics simply because of the frame of mind that you need to be in to create a good building. Are you at peace with yourself? Are you thinking about smell and touch, and what happens when people are walking about in a place? But particularly, are you at peace with yourself? All of that is completely disturbed by the pretentiousness, insistence, and complicatedness of computer graphics and all the allied techniques. So that my final objection to that and other types of methodology is that they actually prevent you from being in the right state of mind to do the design, quite apart now from the question of whether they help in a sort of technical sense, which as I said, I don't think they do.

Jacobson: How and why has design methodology emerged as a special interest area?

Alexander: I think there's a good reason and a bad one. The good reason is that architecture was in a terrible state; is in a terrible state. A lot of people going through school and in their early professional years really couldn't stomach it, were not willing to do what was accepted, and began looking for something that seemed like a more reasonable basis than the really crass fiddling around that some of the architects were doing. That's the good reason. The bad reason, I think, is fear. Plain and simple. It's associated with a psychological state of mind in which a person is not willing to do the rather fearsome thing of creating a design, and backs away from it. I know that that was partly true in my own case when I was interested in the subject. And I think it's becoming more and more true. In other words, even amongst students who are not interested in design methods I find that this fear is very visible; this refusal to commit. The incredible and endless list of excuses as to why we cannot do a design today. And I think in that sense that 'design methods' is just another one of those excuses but, for some people, large enough to excuse them for a lifetime.

I have some more to say on the subject of fear. A book that has just come out called *Emerging Methods in Environmental Design and Planning* asked permission to reprint an article which Barry Poyner and I wrote a number of years ago. I was also asked to make some drawings to go with this article. I made the drawings and they were very rough free-hand sketches. I sent them to the editor, carefully explaining that the roughness and free-handness was deliberate, the reason being very simple: namely, the patterns that I was describing are extremely fluid entities and the free-hand drawing captures the fluidity much better than a precise machine-like drawing. To my amazement, in spite of my request, these drawings had been redrawn in a very unpleasant stiff machine-like way and I looked through the book and realized that there wasn't a single free-hand line in it anywhere. I suspect (I can't prove it) that the people who edited the book, or the press who printed it, or a whole series of people who were associated with the concept of design methods, found it intolerable, unbearable that there should be free-hand lines, free-hand drawings, in this very marvellously pseudo-precise book. I think that's a very serious criticism of the whole thing and it implies to me that there's a state of mind associated with design methods which is really quite nutty and freaky, which is actually something to be watched out for. It may sound trivial. I don't think it is. The idea that the discipline cannot tolerate the idea of a free-hand drawing is a rather serious indication of the state of mind that prevails among the people who practise it. One of the most serious difficulties in the environment today is the machine-like character of buildings that are being made. They are alienating and untouched by human hands. I think it's a horrible state of affairs. I think it's ghastly. I think that people must be able to live in places which have been made by *men*. And any discipline which is so uptight that it can't even tolerate a drawing which was made by a man is almost certainly going to be associated with these kinds of buildings which are *not* made by men. And I won't stand for that.

Jacobson: Are there any problems that design methodology has successfully attacked?

Alexander: There obviously are some problems which have been solved. For instance, there are computer programs which can really help to analyse, and in some cases synthesize, three-dimensional space frames and cable nets. There have always been engineering methods that have helped in the design of structures. Those have become more extensive. I believe that critical path methods help in scheduling jobs. In short, my feeling about methodology is that there are certain mundane problems which it has solved—and I mean really incredibly mundane. The best answer I can give to your question is a personal answer. The fact is that it has solved very few problems for me in my design work. Most of the difficulties of design are not of the computable sort. For two reasons. One is that in most cases design depends on the depth of the insights you have, and any investigation that you want to undertake preliminary to the design has to do with trying to deepen your insights. For instance, in the case of designing a building's lobby, it may very well be that we have a rule of thumb which tells us how big the lobby should be. But the issue at stake is the difference between good lobbies and bad lobbies, and it's quite unlikely that that's going to hinge in any critical way on a very precise determination of the size.

A rather rough determination of the size will usually be quite all right. The difference between a really good lobby and really bad lobby will hinge on much subtler questions which most of us don't know. In so far as we want to study things before or during the design, we want to study what it is that makes a lobby good and that's a problem of insight which is not particularly to be helped by methods. I mean again, of course, when you are studying that you do little experiments, you do all kinds of things to try and help yourself sharpen your own insight. You don't rely on methods in any mechanical sense. The other thing that's going on in design apart from deepening one's insight is the actual fusion of insights to create form. And I do not think this is a particularly mysterious process. What I mean by that is that it's not mystical, it is not beyond discussion. But when you are fusing your insights to create form you're operating in a realm which is so far from the numerical realm, that no method that exists now sheds any useful light on the sort of morphological difficulties you're having while you try to do that.

Jacobson: In what areas should future work centre in design methodology?

Alexander: I think I just have to be consistent here. I would say forget it, forget the whole thing. Period. Until those people who talk about design methods are actually engaged in the problem of creating buildings and actually trying to create buildings, I wouldn't give a penny for their efforts. Anything that somebody says if he is actively trying to make better buildings may be interesting, and I would let that activity itself define the question of what needs to be done because I think once these people get themselves engaged in that activity their notions of what needs to be done will change. And the activity itself will lead them to where they ought to go.

Jacobson: Would it be useful to discover the kinds of difficulties that designers encounter in acts of real design?

Alexander: Well, my own view about that is that these kinds of difficulties have to do with the freedom of the spirit and I really cannot believe that any methodology is going to help that. I do think it's a very serious issue. For instance, we have recently been designing a California Mental Health Center, designing the building on the site with the client. One of the architects associated with us, a practising professional, comes to the site, says he cannot do this—he can only work at the drawing board. Now the fact that he is not free enough in himself to be able, actually to have the nerve to conceive the building right then and there, out on the site with the rest of us, is a difficulty of his. Obviously there are a great many things he could be helped by—but design methodology is not one of them.

Jacobson: Maybe design methodology can identify what kinds of knowledge the designer is lacking at any point. It's not clear for example, what kind of information he lacks.

Alexander: That's right. It obviously is possible to create all kinds of information which would be helpful to people doing design. I think that the patterns that we have developed are very helpful to people doing design and I think there are many other kinds of things which also are helpful. What I shrink from, and in fact reject, is the concept of methodology which I find to be a very barren and intimidating concept. Something that is relatively sensible becomes extremely absurd when you call it methodology. Here is an example. Murray Silverstein and I have been designing a building. At Murray's suggestion, we have been going to an open piece of ground and putting wood stakes in the ground to indicate the organization of the building as we create it, then moving the stakes around and getting the feel of it all. We were talking about this with a friend and she jokingly said, 'Why don't you write an article on the wooden stake methodology, namely, you get yourself some pieces of firewood, you wait for a foggy day, you go out to an old field and you start putting stakes in the ground'. And she said we should write this up and send it in to *Emerging Methods in*

Environmental Design and Planning and submit it as an article. We had a good laugh. It is funny because it is ludicrous to call it methodology. And yet as far as method goes, it is a very serious method which plays an enormously important role in helping to make a better building. When you call it methodology it becomes utterly idiotic and nothing but funny. And in fact I feel that the whole idea of methodology is one step removed from what is real. Anything that is actually real is scorned by people who claim to be methodologists. Anything that is legitimate methodology is accepted precisely because it is so remote from everyday flesh and blood.

Jacobson: What work are you familiar with that would indicate important future directions? Who else is doing interesting things? What about this whole new thing of getting the users involved?

Alexander: Here's a good example of what I was just talking about. I believe passionately in the idea that people should design buildings for themselves. In other words, not only that they should be *involved* in the buildings that are for them but that they should actually help *design* them. I also believe passionately in the importance of information. But the moment these two ideas are brought under the rubric of methodology, I start laughing or crying. It just is nonsense. Why call it methodology? Why be so pretentious? Why does one have to call the simple idea of getting people to design their own buildings—why does that have to be known as a methodology? What's the matter with the people who are calling it that?

Jacobson: I think it becomes a methodology when you make them the actual designers in effect. That is, when you completely step out of the picture and you simply base all decisions on the results of what they've said. In other words when you don't become responsible for it then it clearly is a methodology.

Alexander: But why do you want to call it a methodology? Why not just call it something to do. It's the pretentiousness of the whole thing that annoys me so much. You see this is the point: if you call it 'It's A Good Idea to Do', I like it very much; if you call it a 'Method', I like it but I'm beginning to get turned off; if you call it a 'Methodology', I just don't want to talk about it.

Jacobson: If this is how you feel about design methods, how do you view your earlier book, *Notes on the Synthesis of Form*, and what was your intent then? It is clear that you are viewed as a major theorist in design methods. What is your feeling about that?

Alexander: Well, as far as I am concerned, the whole thing has been a painful and drawn-out misunderstanding. My situation in 1958 was very simple. I wanted to be able to create beautiful buildings. I didn't know how, and nothing that I was learning in school was helping me. Yet at the same time, I had a very clear sense of the difference—I knew what beautiful buildings were—and as far as I was concerned, not only was I incapable of making them, but so were most of the architects then practising. What I wanted to be able to do was to create buildings with the same kind of beauty that traditional architecture had. So I began to find out what to do. This really meant going to the roots of form. To that extent, even the simple emphasis on function, and requirements, in *Notes*, was, for me, merely a way of getting at beauty—a way of getting at the foundations of a well-made, beautiful thing. And the so-called 'method' of that book was, in the same way, simply a process which seemed to me to go to the heart of what had to be going on in a beautiful building.

Jacobson: If this was your intent, why did you present it so clearly and sharply as a 'method'?

Alexander: As you know, I studied mathematics for a long time. What I learned, among other things, was that if you want to specify something precisely, the only way to specify it and be sure that you aren't kidding yourself is to specify a clearly defined step-by-step process which anyone can carry out, for constructing the thing you are trying to specify. In short, if you really understand what a fine piece of architecture is—really, thoroughly understand it—you will be able to specify a step-by-step process which will always lead to the creation of such a thing. Anything short of that means that you don't really understand what is going on. So, for me, the definition of a process, or a method, was just a way of being precise, a way of being sure I wasn't just waffling.

Jacobson: But you did actually use it, didn't you? At what point did you discover that it wasn't necessary to go mechanically through all of the interactions, and then to use the computer programs to get sub-systems, and so on?

Alexander: Well, during my experience in India, designing the Indian village, and then again during the design of the San Francisco rapid transit stations, I began to see that we could go straight to the diagrams for subsystems of forces, without going through the earlier steps of the procedure—and in my later work I began to call these diagrams 'patterns'. But this discovery, in itself, is not essential. That isn't what I want to talk about because there is a danger that people will once again think that what is at stake here is a 'method'—except that it is now a new method, a revised method. That isn't the point at all. The real point concerns the motives behind all this work. My motive, from the very outset, has always been the same: to make better designs. This is a very practical motive. Whenever something doesn't help me make better designs, I get rid of it, fast. What I am most anxious to convey to you, and to the people who read this interview, is the idea that if that is your motive, then what you do will always make sense, and get you somewhere—but that if your motive ever degenerates, and has only to do with method, for its own sake, then it will become desiccated, dried up, and senseless.

5.2 Second-generation Design Methods

Horst W.J. Rittel (Interviewed by Donald P. Grant and Jean-Pierre Protzen)

Grant: What do you see design methodology as trying to do?

Rittel: The occurrence of interest in methodology in a certain field is usually a sign of a crisis within that field. When they talked about methods and methodology in mathematics it was due to the difficulties they had run into with the development of set theory; when the social sciences talked about methods it was when the field was in a crisis. The same is true of the design professions. Important design problems have changed their character from almost professional problems to the type of problem where this approach does not seem satisfactory any more, and therefore they have begun to talk about methodology. The main purpose of design methodology seems to be to clarify the nature of the design activity and of the structure of its problems. This role of design methodology seems to me to be much more important than its practical use in dealing with concrete problems.

Grant: How and why has design methodology emerged as a special interest area?

Rittel: The reason for the emergence of methods in the late fifties and early sixties was the idea that the ways in which the large-scale NASA and military-type technological problems had been approached might profitably be transferred into

Originally published in *The DMG 5th Anniversary Report: DMG Occasional Paper No. 1* (1972), pp. 5–10. Reproduced by permission of The Design Methods Group.

civilian or other design areas. The discovery of the development of the systems approach or mission-oriented approach, as contrasted with the traditional modifying approaches of engineering design, was one of the reasons for the optimism that led to interest in the field.

Protzen: Do you think that these people—the military and NASA people—made some effort to propagate this sort of thing in other fields?

Rittel: Later. But to begin with it was the outsiders who had heard about this and read about this in the emerging literature. I think that in the beginning, outsiders from architecture, engineering, and business heard about the methods of the systems approach and thought that if it were possible to deal with such complicated things as the NASA programmes then why couldn't we deal with a simple thing like a house in the same way? Shouldn't we actually look at every building as a mission-oriented design object?

Protzen: But then doesn't that raise the supposition that these people had a problem that they didn't know how to approach, and thus wanted to apply some new technique?

Rittel: They were dissatisfied with their way of doing things. You could observe this in many areas. It was certainly the case in engineering, where production methods had changed, and they had started to look at the product not as a matter of engineering a single product but as engineering a combination of market and production and servicing and the fit between these things. It was also the case in industrial design when they decided that they should deal not just with cosmetic improvements in engineering hardware but also with the interface between user and object, and it was also the case in architecture. It was later, in the mid-sixties, when the big systems people like NASA were looking for civilian applications in order to have an additional justification for their programmes that they began to believe in the spin-offs of their work into civilian use. Among the technologies that they wanted to transfer was the systems approach, and this is so even today.

Grant: What kinds of problems has design methodology successfully attacked? How important have these successes been to design problem-solving, either in theory or in practice?

Rittel: If you are asking for examples from architectural design I wouldn't know of any building that has been done discernibly better than buildings done in the conventional way. The same may be so in the other fields, although some inventions and developments are claimed to be due to the application of design methods, like the invention of the hovercraft with the help of morphological analysis. The reason for this may be that it takes considerable time before such methods find their practical application within the professions. Another reason may be that the present state of the art in methodology is such that it has little economizing effect on design work—in fact it makes it more involved and time-consuming—and you can get away without applying it in most design fields.

Protzen: Except when working with such agencies that require you to apply such methods as cost–benefit analysis, like Housing and Urban Development.

Rittel: Yes; but what effect that has had, or whether the effects, if any, have been beneficial for the projects involved, I would say cannot yet be determined.

Protzen: Do you see any hope that this could ever be established? That is, whether the application of cost–benefit analysis or any other technique has had a beneficial effect on the overall project?

Rittel: I would not be very optimistic with regard to cost–benefit analysis if it is applied as it has been applied until now. If you see it as a kind of almost objective means of determining what the best of a set of alternative solutions is, and if you pretend that in this computation all costs and benefits to the various affected parties have found their representation, then I would say that it must fail, or that it cannot contribute anything essential to better solutions. But if you would use cost–benefit analysis in a kind of argumentative fashion—that is, by having the proponents of a project and its opponents prepare certain cost–benefit analyses—then I could see a beneficial role for this technique in its stimulation of the discourse evolving among the various parties, as a means of structuring the discussion.

Protzen: Could one ever claim that something got better because of the method?

Rittel: In the long run it doesn't matter how something came about; and what is good in one person's eyes may be very bad in somebody else's eyes. On the other hand one could think of experiments where the same problem is approached in different ways and subjected to the same kind of examination. Then it might be possible to show that the results of method A were better than the results of method B relative to the system of examining the results; but for practical purposes it doesn't matter how something was done, and because every non-experimental problem is essentially unique you can never show or demonstrate how it would have been if you would have generated the solution in some other way. Of course it does matter how you design or make a plan while you are making it. The justification of searching for systematic methods is a certain confidence of hope that they might assist in forgetting less by applying them—even at the expense of a more complicated and time-consuming design process. It is the belief that whenever you think about something systematically and expose it to a kind of organized criticism through a debate or discussion, for example, that the probability of forgetting something essential is not increased; and that belief you may not be able to corroborate or prove through experimental data, at least in real projects, because they're unique and irreversible and so on. However, the manner in which solutions come about does matter in another way; that is that the experience of having participated in a problem makes a difference to those who are affected by the solution. People are more likely to like a solution if they have been involved in its generation; even though it might not make sense otherwise.

Grant: In what areas should future work in design methodology centre? Why?

Rittel: My recommendation would be to emphasize investigations into the understanding of designing as an argumentative process: where to begin to develop settings and rules and procedures for the open-ending of such an argumentative process; how to understand designing as a counterplay of raising issues and dealing with them, which in turn raises new issues, and so on and so on. The reason for this is that there is no professional expertise that is concentrated in the expert's mind, and that the expertise used or needed, or the knowledge needed, in doing a design problem for others is distributed among many people, in particular among those who are likely to become affected by the solution—by the plan—and therefore one should look for methods that help to activate their expertise. Because this expertise is frequently controversial, and because of what can be called 'the symmetry of ignorance'—i.e. there is nobody among all these carriers of knowledge who has a guarantee that his knowledge is superior to any other person's knowledge with regard to the problem at hand—the process should be organized as an argument.

Protzen: I think that is what you understand when you say 'second-generation methods'—that it is not that there are methods of the second generation but that there is an attitude towards planning.

Rittel: It is not only an attitude, it is procedurally different from the first generation.

Protzen: Then the change in attitude calls for different procedures, and these procedures if developed you would call 'second-generation' procedures?

Rittel: Yes. And these methods are characterized by a number of traits, one of them being that the design process is not considered to be a sequence of activities that are pretty well defined and that are carried through one after the other, like 'understand the problem, collect information, analyse information, synthesize, decide', and so on; and another being the insight that you cannot understand the problem without having a concept of the solution in mind; and that you cannot gather information meaningfully unless you have understood the problem but that you cannot understand the problem without information about it—in other words that all the categories of the typical design model of the first generation do not exist any more, and that all those difficulties that these phases are supposed to deal with occur all the time in a fashion which depends on the state of the understanding of the problem. The second feature of the second generation is that it is argumentative, as I explained before. That means that the statements made are systematically challenged in order to expose them to the viewpoints of the different sides, and the structure of the process becomes one of alternating steps on the micro-level; that means the generation of solution specifications toward end statements and subjecting them to discussion of their pros and cons. This process in turn raises questions of a factual nature and questions of a deontic or ought-to-be nature. In the treatment of such factual or deontic questions in the course of dealing with an issue many of the traditional methods of the first generation may become tools, used to support or attack any of the positions taken. You might make a cost–benefit study as an argument against somebody else's deontic statements, or you might use an operations research model in order to support a prediction or argue against somebody's prediction. However, I wouldn't say that the methods are the same just in a different arrangement and with a different attitude, but that there are some methods particular to the second generation, and that these are in particular the rules for structuring arguments, and that these are new, and not in the group of methods developed in the first generation.

Protzen: There are some other concerns that are new, the crucial one being the question of who is to participate in the debate. One could argue that this is not an entirely new area of concern, since operations researchers have long been concerned with who the clients were, but nevertheless I think that this is a new attitude towards design and requires new techniques of determining who the clients are and how they can be drawn into participation.

Rittel: Yes. I would say that these would be methods not in the first generation. First-generation methods seem to start once all the truly difficult questions have been dealt with already.

Grant: Would it be fair to say that a fundamental difference is that in the first generation the difficulties being dealt with were basically technical issues, and in the second generation the basic questions are questions of deontic or ought-to-be statements and of conflicting interests?

Rittel: The second generation deals with difficulties underlying what was taken as input for the methods of the first generation. For example, to set up a measure of performance or an effectiveness function is a focus in the second generation, while in the first generation that was considered an almost trivial task, or at least a task that had already been solved before the procedures to be applied were set in motion; optimization techniques are an example.

Protzen: Wouldn't it be fair to say that in the early writings in operations research, like the Churchman introduction, that in the first chapters they would consider some of these things to be main tasks?

Rittel: Yes, but only for lip service, for while they mentioned these problems in the first chapters, or more likely in the introduction, the instruments they offer don't deal with them; however, we ought to consider things like the Churchman–Ackoff technique and similar ones to be steps in that direction. Another property of the second generation is that upon abandoning the step-by-step structure of the first generation, the classic problem of the first generation disappears: that of implementation of the solution. That is because of the participation of the affected parties; the implementation grows out of the process of generating the solution. The first-generation model works like this: you work with your client to understand the problem; then you withdraw and work out the solution; then you come back to the client and offer it to him, and often run into implementation problems because he doesn't believe you. The conclusion of the second generation is that such a sequence is entirely meaningless, and the client is well advised not to believe you in such circumstances, because at every step in developing such a solution you have deontic or ought-to-be judgments that he may or may not share, but that he cannot read from the finished product offered in your solution. The nightmare of the first generation, implementation difficulties, should disappear or at least be minimised in the second generation; or at least that should be one of the aims. That should be the case from having the clients as accomplices during the generation of the solution.

Protzen: This insight into design problems as being really different from what they were assumed to be in the stepwise processes—did it come to you in the process of applying first-generation techniques, or did you develop it independently?

Rittel: To begin with I became interested in this area because I felt that the methods of the first generation should have some use in other fields than those in which they had been developed, but I got into controversy with the proponents of these methods very soon—1960 or so—because whenever I'd try to use them I'd run into trouble. On the other hand, you could observe that many of the proponents of the first-generation methods, like operations researchers, in some countries at least, tend to withdraw from attacking wicked problems and concentrate on the art of linear programming and queuing theory as objects for their own sake—an academic discipline— and not bother about applicability any more. And of course there are some people who still do this, but I would say that the corporations or other planning institutions who seriously tried to accomplish something with the first-generation planning methods have been disappointed, and that there is a considerable 'hangover' from these methods.

Protzen: The same might be said of the computer.

Rittel: What I have said of methodological software of course also applied to the use of the computer in designing. It is easily seen that design in the sense of forming judgments can never be simulated by a computer, because in order to program that machine you would have to anticipate all potential solutions and make all possible deontic judgments ahead of time before the machine could run. But if you did all that you wouldn't need the computer because you would have had to have thought up all solutions ahead of time. Therefore it is almost ridiculous to claim that there will be a designing machine if design is thought of in this sense. But unfortunately the same kind of optimism with regard to the first generation in design methods included the belief that once you use the computer you will design better. Quite some amount of time, effort, and money has been used to demonstrate that the usefulness of the computer is quite limited in the kinds of concerns dealt with by the second generation.

There should be two areas of emphasis in further work in design methodology. One is the further development and refinement of the argumentative model of the design process, and the study of the logic of the reasoning of the designer. What I mean by logic is the rules of asking questions, generating information, and arriving at judgments. There are a great number of identifiable questions that can be dealt with in this area. The second area of emphasis should be work on practical procedures for implementing the argumentative model: the instrumental versions of the model. Some questions are how to get a group going in an argumentative fashion, how to select the group, and the problems of decision rules.

Protzen: One might specify at this point that these rules do not now exist. No set of procedural rules, such as are applied in legislative bodies, or Roberts' Rules of Order, or any others, really covers this sort of situation.

Rittel: Because they are too coarse to deal with the varieties of entities that you have to distinguish in setting up such a rule system for planning, and therefore there is an urgent demand to think up systems of rules and try them out.

Protzen: Systems of rules how to debate and decide, or should one separate them into one set of rules for debate and another for decision-making?

Rittel: They are somewhat separable, because the emphasis of the second generation is on those parts of the argumentative process that precede formal decision. Argument stops once a formal decision is reached. One of the arts of the second generation is actually postponement of the formal decision in order to enhance the process of forming judgments. In the ideal case rules of formal decision-making wouldn't be necessary at all, because people would become unanimous in the course of discussion. Formal decision has always meant curtailing debate, and therefore the formation of judgments.

And of course there is a third area in importance—that of the technical manner of supporting these procedures. If, for example, you clearly organize a planning process according to such an argumentative model as an IBIS (issue-based information system), you will find that the bureaucratic effort of administering the process is abominable, and therefore one might look for administrative and monitoring computer aids to ease the process.

Protzen: Yes, the development of red tape cutters. And there is the fourth area: apply it.

Rittel: I think the only way to learn something useful about all these foci is through application, and that requires that you look for clients that are willing to go along.

Summary of the characteristics of the second generation in design methodology

Rittel: The first characteristic is the assumption that the expertise is distributed as well as the ignorance about the problem; that both are distributed over all participants, and that nobody has any justification in claiming his knowledge to be superior to anybody else's. Thus there is no logical reason or reason of education for saying, 'I know better than you'. We call this the 'symmetry of ignorance'. The consequence of this assumption is to attempt to develop a maximum of participation in order to activate as much knowledge as possible. This is a non-sentimental argument for participation. It is a logical argument. Do you see that? It's important. There are many sentimental and political arguments in favour of participation, but this is a logical one. Whenever you want to make a sentimental or political case, it's good to use a logical argument.

Then the second characteristic is the argumentative structure of the planning process, i.e. looking at it as a network of issues, with pros and cons. Thus the act of designing consists in making up one's mind in favour of, or against, various positions on each issue.

The third characteristic is that you can always look at a given issue as a symptom of another one. That means that you can work the problem level 'up' to the next level of comprehensiveness, and that this should become a regular part of the discipline, though hopefully not too frequently used. There's a principle of parsimony to applying the principle of raising the level of an issue.

The fourth characteristic of the second generation in design methods is its ideal of the transparence of arguments, because the elemental steps of designing may be judgmental and each additional judgment or deontic question depends on understanding the solution up to that time. You cannot list all the deontic criteria that are to be applied ahead of time, because with every step of the solution the new questions that will come up will be typical of the line of thinking that has brought the solution to that point.

Protzen: Can one add here that this relates to the experimental fact that there is no well-ordered or exhaustive set of deontics?

Rittel: Even if you had a list of all deontics, which of them would dominate another one in a given situation in case of conflict cannot be answered normatively, cannot be judged ahead of time, must be judged in the situation. It is this necessity that underlies the principle that arguments should be transparent.

The fifth principle is the principle of objectification, for the sake of: (a) forgetting less, or reducing the probability of forgetting something that will become important after the fact; and (b) the stimulation of doubt; the more explicitly and bluntly you must state your fundamental objectives, the more readily you are able to cast doubt on them. Another reason for objectification is to increase the probability of raising the right issues, meaning those for which the controversy is greatest, both with regard to the importance of the issue itself and with regard to the divergence of opinion or position on the issue. Thus there are two factors of controversy associated with each issue: its importance and the divergence of opinion on it; weight of importance multiplied by variance of judgment.

The sixth and final principle relates to the control of delegated judgment. If you make a designer or planner or participant spell out what assumptions he has made, then you control his ability to incorporate deontic judgments that the client may not agree with. All kinds of planning are necessarily political, and not merely technical. That seems to be a major difference in assumption between the second and the first generations in design methods.

Perhaps I should add a seventh characteristic, the conspiracy model of planning, that exists to overcome the implementation problem that was mentioned earlier. The implementation problem is only a consequence of the artificial separation between the expert who does the work and the client who has the problem that the work is supposed to deal with. Such an implementation problem naturally vanishes here. The role of the planner in this model is that of a midwife or teacher rather than the role of one who plans for others. Instead, he shows others how to plan for themselves.

Protzen: He might also have the role of keeping the group in motion as it plans for itself.

Rittel: All of which implies a certain modesty; while of course on the other side there is a characteristic of the second generation which is not so modest, that of lack of respect for existing situations and an assumption that nothing has to continue to be the way that it is. That might be expressed in the principle of systematic doubt or something like it. The second-generation designer also is a moderate optimist, in that he refuses to believe that planning is impossible, although his knowledge of the dilemmas of rationality and the dilemmas of planning for others should tell him otherwise, perhaps. But he refuses to believe that planning is impossible, otherwise he would go home. He must also be an activist.

The aim of the second generation is that of self-elimination: the best world is that one that does not need any more planning, without being subject to the maximum of entropy. Or at least the best world would be one where no planning *for others* or on the behalf of others or at others was necessary.

The first generation assumes that there is professional expertise about other people's problems, and that there is an asymmetry of ignorance; that is, that one is justified in saying that he is knowledgeable about another's problem and how it can be dealt with. It assumes that the design process is not argumentative, but that during the first phase the planner sits and listens and understands the problem of the client; and then he thinks; and then the client listens to the planner, and is ill-advised if he doesn't follow his advice. Transparence of argument is not necessary because there is something like a professional ethics guiding the planner, telling him to be objective, detached and so on; and objectification (making understandable) is not necessary because there are objective measures; and a conspiracy model is unethical, because one is a professional; a lawyer, for example, does not conspire with the accused. Rather than modesty there is the expectation of all due respect for professional competence; rather than moderate optimism there is great optimism; rather than an aim of self-elimination there is the aim of getting more involved, so that the system becomes dependent on your services. The more you plan in this way, the more future planning becomes necessary.

Grant: What work are you familiar with that would indicate important future directions?

Rittel: If you had asked me what *developments* are most promising for the future direction of design methodology, I would say it is the increasing discontent with the first 15 or 20 years of belief in the first-generation approach, and in the computer. It is astonishing how slowly that has come about. People like Churchman warned at least 8 or 10 years ago of the consequences of the illegitimate simplifications of the first-generation techniques. But the reaction has been a kind of self-elimination without wanting it. The first-generation design methodology had turned into a sort of academic subculture. In a time of economic recession companies can't afford to maintain operations research departments or computer-aided design departments as symbols of prestige; if it is not paying off, they can't afford to keep it. These are the most promising developments; but that doesn't say anything about work. Work? Well, Churchman is OK. And of course there are all kinds of work showing how not to do it—all those glass bead games at conferences.

What I've observed in the students is very interesting. Within two years they've become very different. All those questions like, 'What has it do with architecture?' have disappeared. They're buying statements that used to be challenged. A few years ago people seemed to get aroused, either because they were on the first-generation side or because they didn't believe in it at all. Now they are openly receptive. It's hard to find any student who opposes our research now; it's like forcing an open door.

5.3 How My Thoughts about Design Methods have Changed During the Years

J. Christopher Jones

HOW MY THOUGHTS ABOUT DESIGN METHODS HAVE CHANGED DURING THE YEARS

You don't know where you are
You don't know which way you're going

IN THE LATE FORTIES I BEGAN TRYING FOR WHAT I'D NOW
CALL A HUMAN FUNCTIONALISM i.e. MAKING DESIGN THOUGHTS
PUBLIC SO THAT THEY ARE NOT LIMITED TO THE EXPERIENCE
OF THE DESIGNER AND CAN INCORPORATE SCIENTIFIC KNOW-
LEDGE OF HUMAN ABILITIES AND LIMITATIONS.

Surely that's one of the most human things in life
To be wrong
We all know that we have more affection for others when
they're wrong than when they're right
Sad, perhaps.
Some people called them systematic methods originally,
They've been mainly rational.

Originally published in *Design Methods and Theories*, **11** (1) (1977), 50–62.
Reproduced by permission of The Design Methods Group.

WHAT I BEGAN TO DO WAS TO RELATE DESIGN THINKING TO
OBJECTIVE OR SCIENTIFIC FACTS ABOUT HUMAN PERFORMANCE.
WHAT IS NOW CALLED ERGONOMICS.
OR HUMAN ENGINEERING.

ALSO I WAS TRYING TO ANTICIPATE NOT IGNORE SIDE-EFFECTS.

This still has not been done, or applied at any large scale outside
the military fields.
But to apply this knowledge to life in general seemed to me
And still seems
An appropriate thing to do.
But it couldn't be done unless the design process
The thinking
The reasons
That led to the lines on the drawing board
Could be spoken about in clear language
Such as a scientist uses when he writes a paper about how
people do this, or how metals do that.
But the design process was illogical and all this knowledge was
logical
So until you had a logical design process
You couldn't join the two together.
Could the car be adapted much more to the people inside it?
We looked in enormous detail at the process of eating
We discovered all the kinds of food and all the actions of eating
The result was nothing like the knives and forks we now use
Our designs looked much more like dental instruments

IN THE FIFTIES I WORKED IN THE ELECTRICAL INDUSTRY TRYING
TO FIT DIALS, CONTROLS, SEATS, CONTROL ROOMS ETC TO THE
HUMAN OPERATORS.

THERE WAS ALSO THE GROWING NEED TO DESIGN LARGER
ENTITIES THAN PRODUCTS: COMPLETE SYSTEMS OF MACHINES,
GROUPS OF BUILDINGS, URBAN CENTERS.

Everyone was coming to realize that all was not well with
technology after all, it was creating problems as well as solving
them.
All the equipment I'd ever handled seemed very badly designed
in that none of it took account of side-effects,
Road accidents for example.
I suppose the trip to the moon is the best example of a complete
system which eventually worked after many failures and which
used this kind of thinking

IN THE SIXTIES THE SITUATION CHANGED

THERE WERE MANY CONFERENCES ON DESIGN METHODS
ENVIRONMENTAL SYSTEMS AND RELATED TOPICS.

I WROTE THE BOOK DESIGN METHODS, A REVIEW OF THE
LITERATURE, AND I BECAME A TEACHER OF DESIGN
AND I BECAME ACTIVE IN WHAT'S NOW CALLED FUTURES
RESEARCH

They all wanted a complete recipe,
It appeals to the Latin mind, especially,
A lot of the design methods literature has been translated into
Spanish and Italian, I'm sure it's no accident that this is so.
There was one student who came to us from Argentina
'When' he said
'Chris, are we going to get the Grand Panoram?'
And this grand panoram didn't exist as far as I knew,
But he expected a complete ideology
Worked out and fully finished.
Many people wanted this and perhaps all students want it all the
time
But I feel one should resist any such thing
If one's to continue living
Other than in the form of a
Intellectually speaking
Or actually speaking

I WAS SEEKING TO RELATE ALL THE DESIGN METHODS TO EACH
OTHER,
AND TO EXPERIENCE.
I FOUND A GREAT SPLIT HAD DEVELOPED BETWEEN INTUITION
AND RATIONALITY, REASON.

THERE WERE BLACK BOX METHODS LIKE SYNECTICS WHICH
WORKED WELL BUT NOBODY KNEW WHY
AND GLASS BOX METHODS, LIKE DECISION THEORY, WHICH
WERE LOGICALLY CLEAR BUT WHICH DIDN'T WORK.

What I could see was this...
What did I find out...?
Well I found,...
In writing it..., ...that book
A standard method of describing each method, so that they
could be compared.
What's striking is that each method begins with a first stage that
is extremely difficult to do
Which has no description of how to do it
Which is intuitive.

WHAT EMERGED
IN WRITING THE BOOK
WAS THAT TO USE DESIGN METHODS ONE NEEDS TO BE ABLE
TO IDENTIFY THE RIGHT VARIABLES
THE IMPORTANT ONES
AND TO ACCEPT INSTABILITY IN THE DESIGN PROBLEM ITSELF

ONE HAS TO TRANSFORM THE PROBLEM AND THE SOLUTION
ALL IN ONE MENTAL ACT OR PROCESS

Where is the essence of the subject?
For me the word in the index with the most sub-entries to it
Is 'Instability of Design Problems'
Which has about ten entries
The whole problem becomes more unstable as you widen it
As you take more and more of life to be part of the problem you
don't get a more stable problem you get a less stable problem.
And this I think is not what the rationalists like.
I think that people who approach this subject because it seems
rational are those who like certainty in life.
If you wish for certainty you might as well leave this subject
alone
Because design is to do with uncertainty
As far as I can see
But a lot of people who do wish for certainty do dabble in it
And I fear they're wrecking the subject

....

Especially transformation became evident...
How to get from divergent thought to convergent thought.

I WAS AT THE TIME DOING DESIGN RESEARCH INTO CAR SEAT
COMFORT, OFFICE DESKS, SLIDE RULES USABLE BY SCHOOL
CHILDREN, AND TRAFFIC AUTOMATION SYSTEMS

Incredible isn't it, the world is full of things, like slide rules,
that college students with an IQ of 120 or more, can use, and
the rest of the population, with a normal IQ of 100, can't use.
Things like telephone handsets, which are used by the public,
are extremely difficult things to design
Because you've got to design them so that a very large
percentage of the whole population can use them.
A very interesting concept this
What percentage of the population can use your design.
You may say, if you're designing a building, that the whole lot
can use it.
But I wonder.
The number of people who are too big or small to use it is quite
high, millions.

Large numbers of old people, children and poor people don't
use telephones
They've never got used to the difficulty.
They may use them for local calls but they can't possibly cope
with complicated calls.
Even though it's an extremely easy thing to use and there are
operators waiting to help you.

IN THE SEVENTIES I REACTED AGAINST DESIGN METHODS.
I DISLIKE THE MACHINE LANGUAGE, THE BEHAVIOURISM, THE
CONTINUAL ATTEMPT TO FIX THE WHOLE OF LIFE INTO A
LOGICAL FRAMEWORK.
ALSO THERE IS THE INFORMATION OVERLOAD WHICH SWAMPS
THE USER OF DESIGN METHODS (IN THE ABSENCE OF COMPUTER
AIDS THAT REALLY DO AID DESIGNING).
I REALIZE NOW THAT RATIONAL AND SCIENTIFIC KNOWLEDGE IS
ESSENTIAL FOR DISCOVERING THE BODILY LIMITS AND ABILITIES
WE ALL SHARE BUT THAT MENTAL PROCESS, THE MIND, IS
DESTROYED IF IT IS ENCASED IN A FIXED FRAME OF REFERENCE.

My experience of design methods in the seventies has been of
writing papers against the subject.
I received a letter asking me to write a paper for the EDRA
conference and instructing me how to do it.
The writer of the letter outlined in ten pages what were the
conditions applying to writers of papers for this conference
Use no more than five pages.
He'd adopted the whole language of behaviourism to describe
to his captive authors exactly how they should do their papers.
And much as I like the man who wrote the letters, I like him
very much really as a person, I still felt pretty annoyed at this
sheaf of instructions he sent out
And I thought to myself "I won't give this paper, I'm damned if
I'm going to ... after all this time in this field to be told by him
... he probably wasn't alive when I started in it ... exactly how
to write a paper."
And I felt how insulting really that we have all this mass of
machine language instruction which we send to each other
In a bureaucratic manner
Throughout life.
Not only for arranging conferences.

....

At that time I suddenly got interested in chance processes
And I discovered the works of John Cage
All about how to use chance to compose music and to compose
lectures
And he says in there

If you receive a nasty instruction
You could refuse it,
You could protest strongly to the person who sent you the
instruction
Or you could do what he John Cage does which is, apparently,
To obey the instruction and then do more.
It's tedious and arduous but a nicer way of responding
So I braced myself a bit
I stomached all the instructions
In five pages exactly.

I composed the paper by chance
It's got characters in it who speak
It's got a character called EDRA
And it's got Walt Whitman, who speaks another kind of
American other than machine language
It's got Kant, who speaks machine language very much better
than we designers do
It's got Jung who says that no method is any good unless the
right person uses it
(which is an ancient Chinese view taken from The Secret of The
Golden Flower)
And it's also got some remarks by Graham Stevens
The man who left architecture as a student in the sixties to
become one of the leading people in inflatables, or blow-ups.
Perhaps you saw him walking on water at the EDRA confer-
ence
In a blown up plastic bag.
I introduced Graham into the paper because his account of how
he did it was so clear-cut.
He said
"We were fooling around with balloons and one day we got the
idea of making them huge and filling them with water, all sorts
of things, and that led to my life in inflatables."
Behaviourism, of the Skinner kind, had taken total command
of the architectural part of design methods by this time and I
still feel that's a mistake.

WHAT I'M DOING NOW IS EXPERIMENTING WITH CHANCE
PROCESSES
(THIS PAPER IS COMPOSED OF A CHANCE MIX OF MY LECTURE
NOTES, IN CAPITALS, AND RANDOM SAMPLES FROM A TAPE
RECORDING OF THE LECTURE, in small letters).
I AM ALSO TRYING TO FIND WAYS OF INCREASING PERCEPTION,
AND OF REDUCING ACTION, IN DESIGN, SO THAT WHAT'S NEW
IS MORE SENSITIVE TO WHAT'S ALREADY THERE.
I'M REALISING THAT IF DESIGNING IS APPLIED TO LIFE, NOT
JUST TO PRODUCTS, SYSTEMS AND SOFTWARE, THERE HAS TO
BE MORE POLITICS (IN THE ATHENIAN SENSE) (TWO-WAY PRO-
CESSES)

AND LESS PLANNING,
IF DESIGNING IS NOT TO BECOME TYRANNY.
I'D LIKE TO TRY OUT EXPERIMENTAL DESIGNS, WITH NO CLIENTS,
AIMED AT RESPONDING TO LIFE AS A WHOLE NOT JUST TO A
SECTION OF IT
AND,
TO COMPENSATE FOR THE EXCLUSIVELY VISUAL AND SPATIAL
QUALITY OF DESIGNING AS IT WAS
I'M TRYING TO LEARN FROM THE TEMPORAL ARTS: POETRY,
FILM, MUSIC AND THEATRE.

SOME CONCLUSIONS

What seems to have happened is this:
—an extension of the scale and influence of design processes, to take account of side-effects and of the integration of products into systems
—a search for greater precision and certainty in knowing what fits human bodies and abilities
—greater uncertainty and instability of design problems as they include more of life
—psychological and social resistance to these effects, resulting in design methods being neglected by the professional designers but flourishing as a rational but useless academic game
—political resistance to the extension of planning, to the fixing of life in a rational mould
—data overload (as yet unhelped by computer aids) which prevents designers applying the vast existing knowledge which could otherwise permit a truly "human functionalism"

Generally design seems to be becoming a social art and to do this properly it seems we need to learn from experimental artists whose happenings and other events are making art a way of living.
Both art and design at last seem like meeting, across the Cartesian split of mind from body, to enable us to find a new genius for collaboration not in the making of products and systems and bureaucracies but in the composing of contexts that include everyone, designers too.
To be a part.
To find how to make all we do and think relate to all we sense and know,
(not merely to attend to fragments of ourselves and our situations).
It was a question of where to put your feet.
It became a matter of choosing the dance
Now its becoming
No full stop

5.4 The Development of Design Methods

Geoffrey Broadbent

Most of the pioneer design methodologists discussed the nature of design as a science before proceeding to their personal descriptions of techniques which, hopefully, designers would be tempted to adopt in practice. And, almost without exception, they took a Cartesian view of designing; breaking the problem down to fragments and solving each of these separately before attempting some grand synthesis.

Each theorist used a different terminology, there were differences in the scale and the level of abstraction at which they treated the parts of a problem, but to quote only the best-known examples, Asimow (1962) with his design *elements*, Jones (1963) with his *factors*, Archer (1963/4) with his *sub-problems* and Alexander (1964) with his *misfit variables* were all clearly trying to apply Cartesian methods in design.

One fundamental tenet of the design science which thus began to emerge was that the designer should abandon, absolutely, any question of preconceived design solutions. Chermayeff and Alexander (1963) had spelt out in some detail just why it was that concepts such as structure, acoustics, and so on carry residues of past attitudes to architecture. It would be necessary to abandon those if one was to take a fresh view of design problems.

There is no doubt that by the early 1970s a new and potentially powerful approach to design *had* emerged, based on analysis, quantification, computer aids, and so on. Horst Rittel (1972) called this 'first-generation' design method. Yet asked to catalogue its achievements, in terms of buildings built, cities designed, and so on, most of its advocates find themselves in difficulties. Of course, there are fragments of design—a

Originally published in *Design Methods and Theories*, **13** (1) (1979), 41–5. Reproduced by permission of The Design Methods Group.

transportation analysis here, an actual building plan there, which *do* owe something to such an approach. But the most striking example of all is usually overlooked because it simply does not *look* like the kind of functionally efficient building which—or so its proponents thought—*should* have been the product of such processes.

More than any other example also it emphasizes the "expert-knows-best" attitude which permeated so much design theory at this time. I am referring to the most carefully *calculated* piece of architectural and urban design that has ever been built; Disney World at Orlando, Florida. Disney's aim in commissioning Disney World was not so much to repeat the commerical success of Disneyland at Annaheim in California as to develop an Experimental Prototype Community of Tomorrow (EPCOT) which would be funded from the profits of his second 'Magic Kingdom'. The whole process of finding a suitable location, buying a tract of land some 14 miles by 7 (27,443 acres) in a carefully planned sequence of small lots—so as not to alert the local community and thus push up prices—of draining the central Florida swamp whilst retaining certain nature conservancy areas—all this was planned with meticulous precision.

Among other things, Disney World represents the most comprehensive application of queuing theory anywhere in the world. On first arrival, having parked one's car, one is picked up almost immediately by a motorized train and transferred to the monorail systems—where a system of ramps, barriers, and chains ensures that waiting passengers are distributed evenly along the station platform *before* the train arrives. From this, through the whole vast system of interrelated 'people movers' to the simple act of queuing for a meal (one walks to a counter, to be served immediately by a Disney-programmed girl, rather than walking *along* a counter and thus being delayed by other hesitant customers), one is conscious of being subject to the most subtle manipulation. There is indeed an Automatic Monitoring and Control System—designed by RCA Victor—by which visitors are constantly monitored (see Haden-Guest, 1972) all of which is consistent with Disney's personal philosophy.

One must remember that Disney's reputation was made by Mickey Mouse and other cartoon animations. Having eventually got them to move and speak as he wanted them to, his interests then extended to real live people. The workers—and the paying consumers—of Disney World indeed are controlled as Disney wanted them to be; EPCOT is merely intended as a further stage in his people-animating process. It is conceived as an integrated Electronic City with a central, 35-storey commerical centre surmounting a Transportation Lobby and surrounded by totally enclosed shopping malls which in turn would be surrounded by theatres, offices, surgeries, and high-density apartments. Disney conceived it with cable television, for instance, screening educational programmes which actually showed people how they 'should' live.

There is much more to it than that, but the crucial point is that in terms of techniques, and more particularly in terms of those attitudes in which the "expert" knows best, Disney World represents the most complete realization of first-generation design methods applied to the built environment anywhere in the world.

Horst Rittel suggested some ten years after the first major publications that this first-generation approach seemed to have died. Its major exponents certainly had withdrawn from the field and suggested, in doing so, that the whole thing was a terrible mistake.

Chris Alexander (1971), of course, had published his elaborate retraction from his former position suggesting that design methods as originally set up actually: 'destroy the frame of mind the designer needs to be in if he is to design good architecture'. He cited an example in which he and his colleagues actually went out on to a site and drove posts into the ground to indicate where the corners of a building would be—as an aid to knowing whether it was the right size or not. He went on to suggest that one could describe this as a design method—the 'post' method, perhaps, but that would sound very pompous. Actually it is the Pragmatic Design which I described in *Design in Architecture* (1973). Much of this disenchantment stems, I think, from the Portsmouth Symposium of 1967. (Broadbent and Ward, 1969.) Tony Ward set this up, in part, as a confrontation between behaviourists such as Markus and Studer, who, in particular, were cast in this role and those who took a Marxist-existentialist view, notably Tony Ward himself and Janet Daley. The behaviourists were characterized as latter-day functionalists wanting to observe human behaviour by empirical methods, to quantify it, to set up models of man/environment interactions and to use these as a basis for designing. The existentialists spoke much more about the individual, responding to his environment with feeling and free to manipulate it as he felt necessary. They drew heavily on the literature which had become required reading in student activist circles, particularly Ronald Laing (1960) implying that the schizophrenic state which he describes so forcibly offers richer and more profound ways of being 'human' than the ordered, linear, one-dimenionsal thinking which the early design methods were designed specifically to foster.

This, no doubt, was old stuff even then to veterans of the Berkeley Free Speech movement. But it pre-dated Nanterre by several months, not to mention the other European manifestations of cultural shift which took place in 1968. This shift, clearly, has had profound repercussions on the pioneers of design method, so we shall have to look at it more closely if we are to put into context what actually remains of design method.

It is hardly surprising that many of these existentialists also subscribed to the Marxist idea of alienation and looked to Marxism to change the society which they found so intolerable. They hoped—with Marx—that capitalism would be over-thrown, yet increasingly it seemed clear, as Marcuse (1964) put it, that capitalism was sapping their creative energy with its 'repressive tolerance'. The workers were content because they could afford and were conned into buying the fruits of their own labour.

Designers had a particular role to play in this; their ingenuity ensured a constant increase in the efficiency of factory produc-tion methods, thus increasing the capital available to capital-ism. But as the production of goods increased so demands for them also had to be stepped up; designers contributed to this by styling new models, intended to make the old ones obsolete. Increasingly also, they associated with, or even became, market researchers, media planners, copy writers for advertising, and so on. Another Jones (1969) called them 'technicians' of *production* and *consumption*, adding also a third group, the 'technicians of *consent*'—journalists, editors, television perso-nalities, film-makers and so on, whose job it was to indoctrin-ate the public with values which make them into willing and contented consumers.

It is hardly surprising that some designers—including some architects—should shrink from prostituting their skills in these ways. They refuse to be a party to *any* activity which inhibits the potential of other people to grow into what *they* conceive themselves to be. So increasingly we find designers who do not *want* to make design decisions, who believe, at most, that their task is to encourage other people to determine what they themselves want. That explains much current interest in citizen-participation (Burke, 1968), advocacy planning (Davi-doff, 1965), and *charette* (Shelton, 1971), in which interested parties are brought, and kept, together, sometimes for weeks if necessary, until they have thrashed out a design solution amongst themselves. Design method seems quite irrelevant in contexts such as these. Or, worse still, it is seen as a 'skill' which the 'expert' will bring to bear in overriding the wishes of those he is supposed to be designing for.

Horst Rittel (1972) suggested that such methods were leading to a second generation of design methods, based on a number of premises, especially:

(1) The assumption that the expertise is distributed ... all over the participants ... nobody has any justification in claiming his knowledge to be superior to anyone else's ... we call this the 'symmetry of ignorance'. ...

and

(2) The argumentative structure of the planning process ... Thus the act of designing consists in making up one's mind in favour of or against various positions on each issue.

He wanted to heal the 'artificial separation between the expert who does the work and the client [whose problem] the work is supposed to deal with'. Rittel's second generation designer therefore is no longer an 'expert' telling people what they should want as much as a 'midwife or teacher' who 'shows others how to plan for themselves'.

At first sight the record of second-generation design methods is somewhat more impressive than that of the first generation. A vast literature on participation indeed has built up, from the first, theoretical statements (Davidoff, 1965) through government legislation such as the (British) Skeffington Report of 1969, through a series of conferences held by the Design Research Society (Manchester, 1971) and with the Design Methods Group (London, 1973) and the American Environmental Design Research Association (EDRA). But theory is one thing and application quite another. In most places where it has been tried, participation actually works in the prevention, or at least the delay, of planning proposals which are going to harm people's interests. Perhaps the most spectacular result so far has been that of the anti-motorway lobby in Great Britain who—after a series of skirmishes with authority at the Archway in London, at Winchester, and elsewhere—actually persuaded the government to delay, and perhaps even to cancel, its entire motorway programme.

But when one looks for actual designs—that is the projection of new building (and planning) forms arising out of participation, they prove to be thin on the ground. The most publicized example worldwide probably has been Lucien Kroll's buildings for the University of Louvain. Kroll got the job because he wanted medical faculty and students to 'participate' in the design. He 'conducted' various groups who were given slabs of coloured plastic representing apartments, students' rooms, dining, social areas, and so on. The groups then shuffled these around on a contoured model of the site and whilst the 'dining' group kept insisting on a separate restaurant block, the others traded-off space within a series of building blocks so that each became an intricate, multi-use structure. One group devoted its attention to circulation routes, measuring distances within the various proposals with pieces of string. There is no doubt that given a framework of this kind—as to what the planning of buildings actually constitutes—non-architects *could* work within it and have great fun doing so. But the framework itself, of course, was set up by an architect, who also determined the division of the building into structural bays and hence the division of the façades into a grid, each square of which could *then* be filled in with the participation of future users and, in certain cases, by the craftsmen who constructed the buildings.

The resultant buildings are amazing collages of rubble, brick, tile, asbestos sheeting, glass, and glass-reinforced plastics, whose random appearance obviously *expresses* the ideal of participation. Visually they are most exciting, but at a more objective level they present a great many problems. The study/bedrooms themselves are inordinately small, the circulation is extremely complex and, above all, the building fabric itself is perversely opposed to any concept of sensible environmental control. One section of La Meme is covered with Miesian curtan wall (is called—for that reason—Les Fascists), but it faces south-west, the worst possible orientation for such a façade in terms of solar heat gain. The famous 'solid' wall of l'Ecole has almost exactly the right amount of glazing for a south-facing façade in these latitudes, yet it actually faces due north! All this results in gross discomfort for those who have to use these buildings. Yet the rich and intricate forms in which they are conceived could have been turned by Kroll himself to maximum environmental advantage—*if* he had possessed and insisted on exercising the necessary expertise. Instead of that, his insistence on total participation—for the best possible motives—has resulted, sadly, in buildings which are *less* acceptable to their users than they could have been if a well-informed architect had exercised his personal skill.

Here, in Erskine's Byker and elsewhere, the community could not participate *until* some vehicle was available for them to participate over. There are very good reasons for this, which show at their clearest in that participationist dream of how people should take responsibility for designing their own environment—the self-build squatter-housing of the Third World. This naturally varies in detail from place to place, but in a typical case the laws of squatting are such that anyone who succeeds in getting a roof over his head between 7 p.m. and 7 a.m. may keep it. So—having collected together bits of wood, cardboard, corrugated iron, asbestos and so on—the squatters take possession on the appointed day and quickly assemble their shacks. They work, of course, according to certain known typologies in terms of the size and shape of the spaces they enclose, methods of construction, and so on.

Some squatters have no further aspirations, they establish a certain lifestyle, equating the work they can find (or want to do) with the resources they need for subsistence. Others see themselves as upwardly mobile and once they have got the necessary resources, begin to 'harden up' their dwellings with concrete floors, brick or hollow-tile walls, corrugated asbestos roofs, and so on. Those with even higher aspirations then cover the walls with stucco, and paint them in pastel colours. They make them look—on a much smaller scale—like the architect-designed villas in the richer parts of the city.

This has many implications. They do not—be it noted—*ever* aspire to build for themselves the kind of multi-storey apartment slabs which government agencies used to think suitable for them. But, given that they have small houses on the ground yet cannot reproduce the traditional house forms from their villages, they simply do not *know* what to do; nor do they have the imagination to see what is possible within the (meagre) resources available to them. So they derive at second-hand from what architects have offered; *try* to do for themselves what architects would have done.

So, in the last analysis, whilst functionalist/behaviourist techniques cannot possibly work, citizen participation, advocacy planning and *charette* cannot work either. At best they may identify a 'highest common factor' of user needs but, compounded by the existentialist designer's needs to become himself, they may mislead him into thinking other people want the same things. Marcuse, after all, *wanted* his workers to revolt, even though they seemed quite content. One wonders how *he* would have defined their needs.

So, both extremes of this particular spectrum—first generation or second generation; behaviourist or Marxist/existentialist—clearly are deceiving themselves. It is quite impossible for either of them to avoid feeding their own preconceptions and values into the solution of design problems.

It is hardly surprising, therefore, that Landau (1965) and Hillier *et al.* (1972) have drawn attention to the parallels which may be drawn between the methodology of science—in Karl Popper's version—and the methodology of design. The scientist, in deciding that certain phenomena are worthy of his investigation, has, according to Popper (1963), also committed himself to them. He will *start* with hunches, guesses, *conjectures* about these phenomena and will tend to collect data which support his conjectures. It will be easy for him, in many cases, to make them self-justifying, but his prime responsibility under the circumstances will be to test his conjectures as rigorously as possible and to disprove them if he can. He should also encourage others to do the same, so that if his conjectures survive all these attempts at refutation, he has a right to hold them, provisionally, as a theory, until a better one comes along. The designer can work by conjecture, as we saw in the case of the functionalists, who actually generated three-dimensional built form on the basis of their preconceptions, whilst, with rare exceptions, the pioneer design methodologists failed conspicuously to prove that their Cartesian methods would actually work to produce real design solutions.

Once we adopt a conjectures and refutations approach we can also admit again that there is no symmetry of ignorance. I may be ignorant of your lifestyle, but if I know my job as a designer I shall at least know more than you do about the technical aspects of your problems. If I use them as the bases for my design conjectures, then because these are based on what I *know* there is some chance, at least, that you will find them acceptable. And if they *seem* to conflict with your lifestyle, then of course, you can always reject them.

It seems to me, therefore, that Rittel's second generation of design methods is now giving way to a third which takes a

Popperian view of designing whilst recognizing that within it there are people, *experts*, whose job it is to make the design conjectures. Their expertise most surely is needed if architecture and planning are to emerge from their present malaise; but unlike their predecessors—from Le Corbusier to Disney—they do not *know* how people should live. They merely offer possibilities which people can take or leave.

Rittel obviously felt that the first generation of design methods was *wrong*. And so, as we have seen, did many of its exponents. The second generation was seen to be right—not to say self-righteous—but that also had its limitations, as we have seen. But the fact is that certain first generation methods actually *work*—and clearly have a useful working life in front of them. And certain second-generation attitudes were based on the best of intentions. Freed from the cant and humbug which permeated the way they were presented, these too are well worth developing and incorporating into the third generation. This also, no doubt, will have its faults and its successes, but its emergence suggests that, in spite of everything, design methods are alive and well.

REFERENCES

Alexander, C. (1964), *Notes on the Synthesis of Form*, Harvard University Press, Cambridge, Mass.

Alexander, C. (1971), 'The state of the art in design methodology', (replies to questions by M. Jacobson), *DMG Newsletter*, March 1971, pp. 3–7.

Archer, L. B. (1963/4), 'Systematic method for designers', *Design*, 1963: April, pp. 64–9; June, pp. 70–3; August, pp. 52–7; November, pp. 68–72; 1964: January, pp. 50–2; May, pp. 60–2; August, pp. 56–9.

Asimow, M. (1962), *Introduction to Design*, Prentice Hall, Englewood Cliffs, New Jersey.

Broadbent, G. (1973), *Design in Architecture*, Wiley and Sons, London.

Broadbent, G., and Ward, A. (1969), *Design Methods in Architecture*, Architectural Association Paper No. 4, Lund Humphries for the Architectural Association, London.

Burke, E. N. (1968), 'Citizen participation strategies', *Journal of the American Institute of Planners*, September.

Chermayeff, S., and Alexander, C. (1963), *Community and Privacy*, Doubleday, New York; Penguin Books, Harmondsworth (1966).

Davidoff, P. (1965), 'Advocacy in planning', *Journal of the American Institute of Planners*, 1 (4), November.

Haden-Guest, A. (1972), 'The pixie-dust papers', in *Down the Programmed Rabbit-Hole*, Hart-Davis, MacGibbon, London.

Hillier, W. R. G., Musgrove, J., and O'Sullivan, P. 'Knowledge and design', in Mitchell, W. J. (ed.) *Environmental Design: Research and Practice*, Proceedings of the EDRA 3/AR8 Conference, Los Angeles, University of California.

Jones, J. C. (1963), 'A method of systematic design', in Jones, J.C., and Thornley, D. (eds) *Conference on Design Methods*, Pergamon, Oxford.

Jones, G. S. (1969), 'The meaning of the student revolt', in Cockburn, A., and Blackburn, R., *Student Power*, Penguin Books, Harmondsworth.

Laing, R. D. (1960), *The Divided Self*, Tavistock Publications (1959) Ltd, London.

Landau, R. (1965), 'Towards a structure for architectural ideas', *Architectural Association Journal*, June 1965, pp. 7–11.

Marcuse, H. (1964), *One Dimensional Man*, Routledge and Kegan Paul, London.

Popper, K. (1963), *Conjectures and Refutations*, Routledge and Kegan Paul, London.

Rittel, H. W. J. (1972), Interview with Grant, D. P. and Protzen, J. P., in *Design Methods Group: Fifth Anniversary Report*. DMG Occasional Paper No. 1, January 1972.

Shelton, A. J. (1971), 'Public participation in practice', *Architects' Journal*, 13 October.

Skeffington, A. (1969), *People and Planning: Report of the Committee of Public Participation in Planning* (Skeffington, A., Chairman), HMSO, London.

5.5 Whatever Became of Design Methodology?

L. Bruce Archer

Design methodology is alive and well, and living under the name of design research.

To tell the truth, I never did like that hybrid expression 'design methodology'. My objection was not only to the corrupt etymology, but also to the impression, conveyed by the term, that the student of design methods was exclusively concerned with procedure. For my own part, the motive for my entering the field (25 years ago, God help me) was essentially ends-directed, not means-oriented. I was concerned to find ways of ensuring that the predominantly qualitative considerations such as comfort and convenience, ethics and beauty, should be as carefully taken into account and as doggedly defensible under attack as predominantly quantitative considerations such as strength, cost, and durability. Moreover, it is demonstrable that the assumptions upon which even the quantitative considerations are based can never be wholly value-free, and I wanted these assumptions to be at least acknowledged in the design process. The study of methods was thus not an end in itself, and was cerainly not motivated by the desire to eliminate or down-grade the qualitative considerations, although a lot of people interpreted it that way.

In retrospect, I can see that I wasted an awful lot of time in trying to bend the methods of operational research and management techniques to design purposes. The earlier check-list type models of the design process, such as that published in *Design* magazine in 1963–4 under the title *Systematic Method*

Originally published in *Design Studies*, **1** (1) (1979), 17–18. Reproduced by permission of Butterworth and Company (Publishers) Ltd.

for Designers turned out to be very helpful to quite a lot of designers, and hardly a week goes by even today without my receiving a request for copies. It went out of print a decade ago. The later mathematical and flow-chart type models, although in many respects less normative, were never accepted by working designers in quite the same way. The reason, I think, is that mathematical or logical models, however correctly they may describe the flexibility, interactiveness, and value-laden structure of the design process, are themselves the product of an alien mode of reasoning. My present belief, formed over the past six years, is that there exists a designerly way of thinking and communicating that is both different from scientific and scholarly ways of thinking and communicating, and as powerful as scientific and scholarly methods of enquiry, when applied to its own kinds of problems.

It is widely accepted, I think, that design problems are characterized by being ill-defined. An ill-defined problem is one in which the requirements, as given, do not contain sufficient information to enable the designer to arrive at a means of meeting those requirements simply by transforming, reducing, optimizing, or superimposing the given information alone. Some of the necessary further information may be discoverable simply by searching for it, some may be generateable by experiment, some may turn out to be statistically variable, some may be vague or unreliable, some may arise from capricious fortune or transitory preference and some may be actually unknowable. In addition, once known, some of the requirements may turn out to be incompatible with one another. As it happens, most of the problems that most people face most of the time in everyday life are ill-defined problems in these terms. Not surprisingly, in the course of evolution, human beings have found quite effective ways of dealing with them. It is these ways of behaving, deeply rooted in human nature, that lie behind design methods.

The first thing to recognize is that 'the problem' in a design problem, like any other ill-defined problem, is not the statement of requirements. Nor is 'the solution' the means ultimately arrived at to meet those requirements. 'The problem' is obscurity about the requirements, the practicability of envisageable provisions and/or misfit between the requirements and the provisions. 'The solution' is a requirement/provision match that contains an acceptably small amount of residual misfit and obscurity. Thus the relationship between design problem and design requirements and design provision lies along one axis and the relationship between design problem and design solution lies along another axis. The design activity is commutative, the designer's attention oscillating between the emerging requirement ideas and the developing provision ideas, as he illuminates obscurity on both sides and reduces misfit between them. One of the features of the early theories of design methods that really disenchanted many practising designers was their directionality and causality and separation of analysis from synthesis, all of which was perceived by the designers as being unnatural.

Another problem was that design theories were so often communicated in language that was alien, too. I do not mean that the wrong kinds of words were used. I mean that words or mathematics or scientific notation alone were themselves inappropriate. I and several of my students and colleagues are putting a lot of energy into examining the proposition that the way designers (and everybody else, for that matter) form images in their mind's eye, manipulating and evaluating ideas before, during, and after externalizing them, constitutes a cognitive system comparable with, but different from, the verbal language system. Indeed, we believe that human beings have an innate capacity for cognitive modelling, and its expression through sketching, drawing, construction, acting out, and so on, that is fundamental to thought and reasoning as is the human capacity for language. Thus design activity is not only a distinctive process, comparable with but different from scientific and scholarly processes, but also operates through a medium, called modelling, that is comparable with but different from language and notation. Moreover, modelling in various forms, covert or externalized, constitutes the vehicle for all sorts of other activities, not normally associated with design, such as navigating, surgery, dancing, and even crossing a busy road. [....]

Where does this leave design methodology?

Design methodology is alive and well, and living in the bosom of its family: design history, design philosophy, design criticism, design epistemology, design modelling, design measurement, design management, and design education.

Name Index

Subject Index